El Niño: fact and fiction

El Niño

Fact and fiction

BRUNO VOITURIEZ GUY JACQUES

Translated from the French by Ray C. Griffiths

IOC Ocean Forum Series | UNESCO Publishing

Acknowledgements

We wish to thank the scientists of the Institut de Recherche pour le Développement (IRD, formerly ORSTOM) who helped us at different stages of the drafting of this work: Philippe Cury, Philippe Hisard, Luc Ortlieb, Yves du Penhoat and Joël Picaut. This book owes much to their healthy pragmatism which, thanks to their very deep knowledge of the field, allowed them to confront critically the theoretical approach with the reality of the data.

Published in 2000 by the United Nations Educational, Scientific and Cultural Organization
7, place de Fontenoy, 75352 Paris 07 SP

Composed by Nicole Valentin
Printed by Jouve, 18, rue Saint-Denis, 75027 Paris Cedex 01

ISBN 92-3-103649-1

Preface

Without question, the 1997–98 El Niño became a major news story, even a household word. Reports of floods, forest fires, droughts, fishery collapses, along with the ensuing loss of lives and property, regularly featured in newspapers and magazines all over the world. Many people thought that this was a new phenomenon, an emerging menace related to climate change. But we know that this is not the case. El Niño is an ancient, recurring climatic event on our planet. In contrast to the precise recurrence of the seasons, El Niño is irregular and multiannual and this may help to explain the difficulty that many people have in comprehending its vagaries.

It is a matter for serious reflection that society is still asking questions and seeking answers which can explain the many and varied catastrophes associated with this phenomenon. This is so, even after several decades of increasing scientific knowledge of El Niño, mostly achieved through international co-operation in the fields of oceanographic and climate research.

The communication of research findings is important not only in terms of improving public understanding of science, but also in terms of improving public policy-making. The links between the way we produce and use scientific knowledge constitute an interesting sociological phenomenon. The critical question here is: How much of the available knowledge is actually used by the political establishment in its decision-making?

UNESCO, through its Intergovernmental Oceanographic Commission, has been intensively involved in the development of the knowledge base and the observations needed to explain this and other natural phenomena and to advise the Member States on possible means of mitigating their impacts. What has been achieved is only a fraction of what we still need to develop, if we want to help make the continued use of the ocean fully sustainable.

In order to succeed in this endeavour, we need to give full support to institutions responsible for producing and delivering scientific advice to decision-makers everywhere. Equally, we need to respond to public demand for information. This publication is an attempt to contribute to meeting these needs, with regard to El Niño.

Contents

Foreword

In the Andean highlands, at least twice every decade, unusually heavy rains inundate the upland valleys, increasing the risk of mud slides, the much feared '*huaicos*' (Quechuan, pronounced 'whykoss'), bringing a trail of death and destruction. Long before we became accustomed to the modern name of 'El Niño', the Inca civilization, high in the mountains of South America, had successfully adapted to this recurring natural phenomenon. In an extremely arid terrain in northern Peru, the Incas, piling stone upon stone, built an extensive network of aqueducts. At certain points of the network, the walls were not made of stones but of interwoven canes. When these weak links gave way to floods due to El Niño, alternative channels redirected the excess flow down the side of the mountain. These safety valves prevented the destruction of the ancient engineering works during such floods. Notwithstanding the efforts by the Peruvian Government to use this past experience during the 1997–98 El Niño, by creating two artificial lakes from the excess precipitation, most modern societies still have not learned how to adapt to this ancient climatic phenomenon.

Today, we know that El Niño is a feature of a global climatic phenomenon. It is characterized by the displacement of major weather systems from their normal locations. Places accustomed to wet weather become dry and deserts are flooded, with a range of ensuing and related consequences for local agriculture, fisheries and human health. Although it is a global phenomenon, its effects are local and diverse, and not all of them can be characterized as bad or catastrophic.

Bruno Voituriez and Guy Jacques have written a lively account of El Niño, its dynamics and practical consequences. Without sacrificing rigour they have succeeded in introducing the reader to the complexities and limitations that science faces in its quest to understand the climate system. By telling us about the limits to the predictability of highly complex non-linear climate systems such as El Niño, they take us to the front door of chaos theory, one of the most recent developments of modern mathematics.

The ocean plays a fundamental role by triggering and influencing this type of climate anomaly. To understand climate, the vagaries of weather on different time and spatial scales, we need to study and understand the ocean. Research on El Niño opened a wide road to the integration of oceanography and meteorology. This has culminated in the development of a Global Ocean Observing System (GOOS) and a Global Climate Observing System (GCOS), with a high degree of interaction and co-ordination between them. Both are, at present, based on three predecessors: the Integrated Global Ocean Services System (IGOSS), the Ships-of-Opportunity Programme (SOOP) and the Data Buoy Co-operation Panel (DBCP). The underlying objectives are to monitor sea level and real-time changes in key ocean parameters. These observing systems are being developed in light of the results from the World Climate Research Programme (WCRP), which is conducted under the joint auspices of the World Meteorological Organization (WMO), the International Council of Science (ICSU) and UNESCO's Intergovernmental Oceanographic Commission (IOC).

The ability to forecast El Niño has been greatly enhanced by the establishment of an observation system operating in the equatorial Pacific Ocean. Seventy moored instrument platforms have been anchored on the ocean floor and are held in place by a floating buoy filled with oceanographic and meteorological instruments, a true miniature artificial island. Some of these instruments sample the ocean down to a depth of 2,000 metres; they caught the first unequivocal signal of the birth of the last (1997–98) El Niño. This system, the Tropical Atmosphere–Ocean (TAO) array, is now part of the Global Ocean Observing System (GOOS).

After thirty years of fruitful joint work, we are in a position to extend our forecast capacity to ocean phenomena on different time and spatial scales. In a simple analogy with a radio dial, studying El Niño events that occur with a periodicity of between three and five years is like tuning into one or two radio stations. Today GOOS is developing the observing capacity that should enable us to tune in to all the radio stations on the ocean dial. In doing this, the Intergovernmental Oceanographic Commission is fulfilling its mission to help develop the knowledge base and the observations needed for the utilization of the ocean.

I wish to express my appreciation to Fugro Global Environmental and Ocean Sciences Ltd, our co-sponsor, for their support of the Commission's efforts to provide, through the IOC Ocean Forum Series, important contributions to the literature on current ocean issues for the international community. Such public-spirited gestures are very helpful in accomplishing our educational goals.

For this English version of the book, the Commission also received the co-sponsorship of the International Research Institute for Climate Prediction (IRI), at the Lamont Doherty Earth Observatory, Columbia University. I therefore take this opportunity to thank the IRI for its most welcome support for the present effort in the development of the IOC–WMO–UNEP Global Ocean Observing System.

Patricio A. Bernal
Executive Secretary
Intergovernmental Oceanographic Commission of UNESCO

Understanding and forecasting the oceans

Media descriptions of the impact of the recent El Niño event have heightened public awareness of the degree to which atmospheric and ocean climate are inextricably intertwined at local and global scales. This has resulted in an increased public perception of the fundamental practical importance of understanding the marine environment and predicting ocean climate.

Throughout history, knowledge of the behaviour of the world's oceans and seas has played a critical role in human affairs. Without marine information it would have been impossible to open up world trade, conduct naval warfare, protect against coastal flooding or utilize marine resources.

In more recent times there has been a growing recognition that human activity can result in unacceptable damage to the marine environment. Once regarded as a boundless resource, oceans and seas increasingly require management. This has driven a rapidly growing demand for marine information to support effective management and protection of the marine environment.

In the present millennium, demands on the marine environment are set to grow rapidly. The safe and sustainable exploitation and use of marine resources and the safeguarding of both local and global environments will increasingly depend upon our capacity to better understand and forecast the behaviour of the oceans and the atmosphere.

As a leading worldwide provider of meteorological and oceanographic observation and forecasting services, Global Environmental and Ocean Sciences is engaged in the business of linking scientific understanding to the solution of practical maritime engineering and marine environmental protection problems.

Much of our work is concerned with effectively communicating oceanographic and meteorological knowledge to specialists in other fields. We are therefore very pleased to be associated with a book that seeks to improve non-specialist understanding of El Niño as an oceanographic and meteorological phenomenon and to promote a wider perception of the benefits of further research into ocean behaviour and ocean forecasting.

Ralph Rayner
Managing Director
Fugro Global Environmental and Ocean Sciences Ltd

The four main GEOS offices:

Fugro GEOS Limited
Gemini House
Hargreaves Road
Swindon, SN2 5AZ
United Kingdom
Tel: (+44) 1793 725766
Fax: (+44) 1793 706604
e-mail: geosuk@geos.com

Fugro GEOS Limited
Southampton Oceanography Centre
Empress Dock
Southampton, SO14 3ZH
United Kingdom
Tel: (+44) 1703 596009
Fax: (+44) 1703 596509
e-mail: seadata@geos.com

Fugro GEOS Incorporated
P.O. Box 740010
6100 Hillcroft (77081)
Houston
TX 77274
USA
Tel: (+1) 713 773 5699
Fax: (+1) 713 773 5909
e-mail: geosusa@geos.com

Fugro GEOS Private Limited
Box 5187, Loyang Crescent
Singapore 508988
Tel: (+65) 543 4404
Fax: (+65) 543 4454
e-mail: geossingapore@geos.com

Also, visit our website:
http://www.geos.co.uk

The oceans and climate prediction

In the late 1980s, scientists around the world began to recognize the enormous impacts of the El Niño – Southern Oscillation (ENSO). These effects are all the more critical in the developing countries.

Advances in ocean and atmosphere observing systems, the theoretical understanding of the ENSO phenomenon, and modelling of the coupled ocean–atmosphere system led to the idea of establishing the International Research Institute for Climate Prediction (IRI). The concept of the IRI emerged from the growing awareness of the scientist's responsibility not only to improve climate prediction, but also to make the knowledge gained from prediction readily available to decision-makers in affected countries. These ideas have grown out of the scientific discussions held during the implementation of the international Tropical Ocean and Global Atmosphere (TOGA) study in 1985–94.

By the early 1990s, IRI pilot efforts in experimental forecasting and in training were under way. The successes of these programmes led, in the same decade, to the development of the IRI as an institute with a mission to foster the improvement, production and use of global forecasts of seasonal-to-interannual climate variability for the explicit benefit of society.

With a regular prediction system now in place, the Institute enters the new millennium with a focus on real applications of climate-prediction information. Projects and programmes are starting in areas of the world where predictable climate variability significantly affects society, particularly the more vulnerable developing world.

From its earliest days, the IRI has been involved in the Integrated Global Ocean Services System, an IOC–WMO programme, maintaining an extensive user-friendly data library (see http://iri.ldeo.columbia.edu). In addition, senior IRI scientists were involved in the scientific design and implementation of PIRATA (Pilot Moored Research Array in the Tropical Atlantic), a programme affiliated with the IOC–WMO–UNEP Global Ocean Observing System, because of the role of the world ocean, particularly the tropical Atlantic Ocean, with a view to establishing an 'end-to-end' climate-prediction system.

IRI is therefore very pleased to support the Intergovernmental Oceanographic Commission of UNESCO in the publication of this book.

Antonio Divino Moura
Director, IRI
Lamont Doherty Earth Observatory
Columbia University
61 Route 9W, Palisades
NY 10964–8000
USA

"El Niño comes running up from Easter Island, tepid and sickly, the offspring of death by water, beating against the Peruvian coast, suffocating the anchovies and algae in its hot embrace, kidnapping the vital equatorial nitrates and phosphates, breaking the vast food chain as well as the procreation of the great sea fish: heavy and sweating El Niño swims, hurling dead fish against the walls of the continent, stupefying and putrefying it all, water sinking water, the ocean asphyxiated in its own dead tide, the cold ocean drowned by the hot ocean, the winds driven mad and pushed off course. Destructive and criminal, El Niño flattens the coasts of California, dries out the plains of Australia, floods the Ecuadorian lowlands with mud."

Carlos Fuentes, *Christopher Unborn*
Farrar, Straus & Giroux, LLC
Book Publishers, New York, 1989

Introduction

Prior to the Winter Olympic Games in Nagano in 1998, which were greatly at risk of being affected by excessive snowfall, Mr Kobayashi, Director of the Organizing Committee, stated: 'Once the problem of the downhill is solved, only El Niño will stop me from sleeping.' Luckily for him, the games went off successfully; but what is this malicious El Niño thus thought of as wanting to sabotage the games?

In 1997–98, El Niño was regularly in the front-page headlines of the media. It was held responsible for the flooding in the Andean countries and in California, for the drought, accompanied by enormous forest fires, in Indonesia, which everybody could see thanks to the satellite images of the spectacular smoke plume extending from Malaysia to the Philippines. The drought in Brazil, South Africa and Zimbabwe, the floods in Kenya, the typhoons in Polynesia, etc., were also attributed to it. No calamity seemed able to escape the all-powerful El Niño or its sister La Niña, more recently recognized and the antithesis of El Niño, but just as dangerous. The increased cyclone activity in the Caribbean (Hurricane Mitch in the summer) and the flooding in China were attributed to it in 1998.

There was something irrational in the manner in which the media, sometimes with the complicity of scientists, exploited this personification of a natural phenomenon. Even if its consequences are often disastrous, they are also sometimes beneficial, and it is after all just one climatic perturbation among others, which develops on a time scale (from a season to several years) to which humankind is particularly sensitive. At a colloquium held at UNESCO in 1999, a Nobel prize-winner in physics declared that science, far from providing certitude, was killing myths while creating mysteries. He could have added that the mysteries engender the myths and that science was contributing, reluctantly, to the creation of new myths. We can ask ourselves whether the infatuation of the media with the El Niño/La Niña couple springs from mythical thinking, translating the fact that, for most of our contemporaries, the changes in the weather retain their mystery in spite of the efforts of the meteorologists and climatologists to predict and explain. According to the press, El Niño could be an incarnation of – or a substitute for – Tlaloc, the Aztec god, who made rain or fine weather whenever he wanted to. Caprice of the heavens? Caprice of the gods! The idea that the human race has long formed of the changes in the weather and of the climate can be summarized in this way. The most universal illustration is undoubtedly the Flood, a radical way for the Creator to rid himself of a creature that no longer had the good fortune to please him. There is doubtless still, more or less suppressed, something of this perception in the idea we have of the caprices of the weather to which we are subject, and in our persistent scepticism about weather forecasting.

Yet, in recent decades, our knowledge of the climate system and our forecasting capability have advanced considerably thanks to international research programmes carried out since the 1970s. But any forecast carries a risk of error which increases the further into the future the forecast is made. After the scientists had been accused of a narrow and totalitarian scientism, just when they were giving up the idea of announcing certitudes, this assurance, for which they were reproached in the past, is now held against them! We have to learn to live with the uncertainty that progress in research is steadily reducing, even as it pushes back the forecasting horizon, without ever quite succeeding in eliminating the uncertainty.

The 'media reputation' of El Niño results from scientific advances that demonstrate the fact that physical relations exist between the climatic perturbations of regions as far apart as Kenya and the north-eastern United States, or between the Indian monsoon and rainfall in Peru. El Niño is evidently not the *diabolus ex machina* of climate variation, but it crystallizes this essential discovery of the relations amongst the climatic perturbations of various tropical regions, primarily, but also with the rest of the world. This is an important step forwards that allows us to improve the predictive climate models and to test their performances, since the verification follows the prediction by only a few months. As a manifestation, over a time period of a few years, of the variations of a climate system that occur on all time scales, a deeper knowledge of the El Niño/La Nina couple leaves hope of progress in the forecasting of climate fluctuations and their long-term evolution in the face of another threat: the increasing concentration of greenhouse gases in the atmosphere.

It is the story and the results of this scientific adventure that are reported in this book.

1 Why does the climate vary?

A GLANCE AT PAST CLIMATES

The myth of an earthly paradise destroyed by the harmful action of human beings dies hard, whether we evoke a hypothetical golden age or a natural equilibrium dear to certain ecologists for whom everything would be for the best in the best of all possible worlds if it were not for the presence of human beings. But there is no natural equilibrium: the climate, the ecosystems are constantly evolving, whether human beings are present or not. For more than a million years, the Earth has been varying between hot periods, as at present, and glacial periods during which the average temperature is lower by three to four degrees.

The last Ice Age goes back 20,000 years. Glaciers like the one now in the Antarctic covered Scandinavia, northern Germany, Canada and the northern United States. The inhabitants of western Europe – excellent cave decorators – experienced a climate and an environment similar to the ones in present-day Scandinavia. There is nothing to indicate that the Earth will not undergo another Ice Age in the next few thousand years.

'Optimum climatic conditions' – warm periods, such as we now have – are brief. Optimists think that the accumulation of greenhouse gases leading to global warming has come just in time to save humanity from having to face the dread of a new Ice Age. Who knows?

These oscillations between Ice Ages and interglacial periods are recent in the Earth's history (last 2 to 3 million years) and are typical of the Quaternary Period. Going back much further, to the Cretaceous Period (between 135 and 65 million years ago), the Earth was totally free of glaciers: the temperature of the air was about 5°C higher than it is now, and the temperature of the deep ocean did not go below 10°C, whereas now it is close to zero. In the mid-Palaeocene Epoch, 57 million years ago, the temperature on the coasts of Antarctica approached 20°C in summer. The polar ice-cap on this continent only started to form 10 to 15 million years ago.

Closer to home, between the 16th and 19th centuries, Europe experienced a relatively cold period – the 'Little Ice Age' in which the temperature was more than 1 degree lower than it now is. On 26 June 1675, the Marquise de Sévigné wrote to her daughter, who was staying in Provence: 'It is terribly cold. We are keeping ourselves warm, as you must be too, which is an even greater miracle.' And she adds on 24 July: 'We are experiencing a strange kind of coldness. The course of the Sun and the seasons is quite abnormal.' The River Thames was frequently frozen in winter to a depth of 20 centimetres, which allowed fairs to take place on it between 1607 and 1813; this and the retreat from Russia in 1812–13, during which Napoleon's army lost 400,000 men, are two other events of this period.

In contrast, in the 10th and 11th centuries, more clement climatic conditions allowed the Scandinavians to set up colonies in Greenland and in Vineland (North America). This colonization lasted until the climate cooled again, making navigation between Iceland and Greenland very dangerous. Further back in time, at the end of the last glacial period, 8,000 to 10,000 years ago, the Sahara underwent a rainy period, and the Amazonian forest, the symbol of present-day ecology, still mistakenly referred to as the 'lungs of the planet', was restricted to a few small 'islands' in a vast savannah.

THE CLIMATE MACHINE

The climate system is a machine that converts and distributes the solar energy that the Earth absorbs – about 240 watts per square metre. This input is about 10,000 times the world's production of heat and electricity. Part of it (30%) is reflected back into space by the atmosphere, so is lost to the climate system. Another part (20%) is absorbed by the atmosphere which is thus heated. The remaining 50% reaches the Earth's surface and its biosphere and is absorbed: 18% by the continents; 32% by the oceans. The latter fraction plays an essential role in the regulation of the climate, since the land and especially the oceans return part of this absorbed energy back into the atmosphere:

- by *radiation*, since all bodies have a characteristic radiation related to their temperature. The solar radiation received by the Earth is typical of a body with a temperature of 6,000°C. The surface of the Earth, which has an average temperature of 15°C, radiates at infra-red wavelengths; this radiation is absorbed by the atmosphere – the natural greenhouse effect;
- by *conduction*, that is, direct transfer of heat between two bodies in contact, from the hotter one to the cooler one;
- by *evaporation*, which is the most important factor; it gives the ocean a preponderant role, since, as a result of the evaporation, the ocean cools and the atmosphere gains the corresponding energy when the water vapour condenses.

The atmosphere is thus mainly 'heated from below'. It is in the tropical regions that the ocean and the continents receive the most solar energy and that the ocean is at its most 'generous' with the atmosphere; the tropical ocean is the climate system's 'boiler room'.

Thus supplied, the atmosphere is set in motion and transfers the received energy towards the coldest regions at high latitudes. Owing to the Earth's rotation, this transfer is effected in the form of eddies of various sizes: anticyclones, depressions, cyclones and related phenomena. Ocean currents are also born out of this uneven distribution of heat energy and take the form of large gyres that transfer polewards as much heat as the atmosphere does. The motion of these currents is accelerated by wind action, which is a way for the atmosphere to restore, in mechanical form, a part of the energy gained from the ocean by evaporation. The Gulf Stream, which forms the western side of a vast ocean gyre, is a manifestation of heat transfer by the ocean.

There is thus a high degree of coupling between the ocean and the atmosphere to distribute solar energy over the planet and to ensure the functioning of the climate system. The continents and their vegetation also have their role to play by exchanging their energy with the atmosphere, but in a more limited and static way, which is insignificant on the time scale we are concerned with. The cryosphere (the polar ice-caps, sea ice, glaciers) also has an important part to play in climate dynamics. Owing to its strong reflecting capacity, it sends most of the energy received back to the atmosphere. The greater the frozen surface, the less heat the Earth absorbs. The cryosphere is also a tremendous freshwater reservoir; during glacial periods, the sea level is about 120 metres lower than it is now.

VARIATIONS IN THE CLIMATE SYSTEM

The 'climate machine' is, then, a complex system with several components: the Sun, the Earth's orbit around the Sun, the continents, the ocean, the atmosphere, the cryosphere and the biosphere. All these components are constantly evolving at their own speeds. Any variation or disturbance that any one of these components undergoes will have an impact on all the others, which will in turn react at their own pace. The climate system strives to attain an equilibrium it can never reach.

Let us take an example: the Earth rotates around the Sun in an ellipse, of which the shape and position vary. Furthermore, energy received from the Sun and its distribution over the Earth fluctuate over a time scale of 10,000 to 100,000 years, which explains the successive glacial and interglacial periods.

Another example over an even longer time scale concerns tectonic plate movements, which are not unrelated to the particularly hot period of the Cretaceous and its subsequent evolution. Indeed, the distribution of the continents around the globe considerably modifies the transport of heat by sea currents and their exchanges with the atmosphere.

On a time scale closer to human concerns, we are well aware of the seasonal cycle. We worry about its variations and should like to be able to predict them. At the same time, we ask ourselves questions on the impact of increasing greenhouse gases on the climate of the twenty-first century. In fact, from the length of one season to millions of years, a whole series of causes of variability over different periods of time have intermingled, making any equilibrium impossible and making climate predictions very difficult. On the time scale that interests us, it is the ocean–atmosphere coupling that is the main regulator of climatic variation and is worth more of our attention.

THE ATMOSPHERE

It may seem pointless to worry about the climate of the next few months or years when we are constantly faced with the limitations of weather forecasts. At present, the meteorological services can make predictions up to seven days in advance. It seems that it will be impossible to make predictions beyond fifteen days. This prediction is based on models constructed according to the laws of physics that govern atmospheric dynamics. Starting from the state of the atmosphere at a particular moment, based on data collected by meteorological stations around the world and by Earth-observing satellites, the model calculates the state of the atmosphere, hence the weather, one, three or seven days later. The prediction thus combines observations and models, but it is almost certain that the state of the atmosphere at any given moment is independent of what it was fifteen days earlier. Moreover, whatever the accuracy of the observations and the models, the current situation tells us nothing about what it will be fifteen days later. Such a prediction is impossible. To recall what the meteorologist Edward Lorenz said on what is known as the 'butterfly' effect: the flutter of a butterfly's wing in China could be responsible for a cyclone in the Antilles a few days later. To put it differently, a prediction beyond fifteen days would necessitate knowing the exact state of the atmosphere at all points around the world, to a precision equivalent to the fluttering of a butterfly's wing, which is strictly impossible. In other words, the atmosphere has no memory, and its current information is completely dissipated within fifteen days.

THE OCEAN

The ocean has a slower rate of evolution and therefore a much better 'memory'. It has two main roles: supplying part of its energy to the atmosphere, and distributing the remainder directly around the globe via its currents. In any given place, the amount of energy exchange with the atmosphere depends on the surface temperature of the ocean, and therefore on the amount of heat that has been transported thereto. The part of the ocean to be taken into consideration in climate processes depends on the time scale chosen. If we simply require meteorological predictions of less than two weeks, the models only need the sea-surface temperature to determine the heat-energy exchange between the ocean and the atmosphere. During this time, the sea-surface temperature changes are too small to have a significant impact on these exchanges; it would be pointless to complicate the models by introducing ocean dynamics.

However, on climatic time scales, it is the slower component, the ocean, that dictates climate variability. For monthly or interannual changes (El Niño, for example), it is the first few hundred metres of the equatorial zone of the ocean that play a vital role. Beyond this time scale, the entire oceanic circulation from the surface to the sea bed has to be considered; this cycle takes place over several centuries. The ocean can bear the 'signatures' of previous climatic events for hundreds of years. The present climate depends to a certain extent on the cooling of the Earth during the Little Ice Age mentioned previously. Even if the ocean absorbs the impact of climate variations, it can also restore their effects decades or even centuries later.

EL NIÑO

The tropical oceans are the main suppliers of the atmosphere's energy. The largest of them, the Pacific, the width of which, at the equator, covers almost half the circumference of the globe, plays a predominant role in climate regulation. Any disturbance in its energy exchange with the atmosphere has significant repercussions on the Earth's climate.

El Niño is the tangible manifestation of the interannual variability (from a season to a few years) of the ocean–atmosphere coupling in the tropical region of the Pacific; but it is at this level that human beings are the most vulnerable.

The fact that we can, a priori, link meteorological events, ranging from floods in Peru and East Africa, to droughts in Australia, India and Brazil, and a decrease in the number of cyclones in the Caribbean, to one phenomenon that we are just starting to understand and to be able to predict, constitutes a vital discovery giving hope for climate prediction. This explains El Niño's 'success' with the media; it has been portrayed as an evil magician, whereas it is simply a variable natural component of the climate. For researchers, it is an ideal case study to test the possibility of predicting the climate and to verify the validity of the models, since the effect, hence the experimental verification, closely follows prediction. But before reaching this stage there was a long tale of events which deserves to be told.

2 The story of an encounter between the ocean and the atmosphere

EL NIÑO: BONANZA OR CATASTROPHE?

For the past few decades, the El Niño phenomenon has been covered by the media and given more and more importance each time it occurs. Such attention seems to indicate that the phenomenon is getting worse and inevitably reminds us of another climate 'headline': the increasing greenhouse effect, of which one of the first consequences could be, in fact, an intensification of El Niño. From this point of view, the years 1997–98 were particularly interesting, since there was a link between the two phenomena. Just when microphones and cameras were all fixed on Kyoto where an international conference on climate change was being held, with the aim of reducing greenhouse-gas emissions, El Niño 1997–98, presented as the El Niño of the century, was reaching its peak, having left a long trail of catastrophes in its wake: droughts and fires in Indonesia, floods in South America and in the Horn of Africa (Somalia).

THE FISHERMEN OF PAITA – OR EL NIÑO OF THE FIRST KIND

El Niño is the Spanish name lovingly given to the infant Jesus (El Niño Dios). It is therefore somewhat blasphemous for Christians to associate their Saviour's name with the avalanche of catastrophes related to it. How did we arrive at such an unnatural association?

El Niño entered the scientific realm in 1891 thanks to the then young

Lima Geographical Society. Several of its members were impressed by the extent of the rainy season: from February to April, torrential rains poured down on the desert coastal region of northern Peru, causing havoc in the urban centre of Piura and the port of Paita. The bridge over the River Piura, which was built in 1870 and had resisted four major floods, was swept away by the exceptional flood in 1891. These geographers were the first to make a link between these exceptional downpours and the simultaneous presence of abnormally warm waters along the coast. Brought by an ocean current running from north to south, these waters are easily identified by the debris that they carry from the Gulf of Guayaquil: palm leaves, bananas, tree trunks, dead alligators, etc. Captain Camilo Carrillo, an experienced sailor, linked these observations to a coastal current known to the fishermen of Paita as *el corriente del Niño,* simply because this weak and small stretch of current appeared almost every year at Christmas time. It coincided with the rainy season, which is most welcome in such an arid region, especially for husbandry and cotton growing. The fishermen themselves, who practise small-scale fishing, benefit a lot from this warm current, since they then have access to prized tropical species: dolphinfish, yellowfin tuna and bonito, octopus, shrimp, etc. For them, El Niño is a blessing, or in primitive terms, it is Father Christmas! Sometimes, however, its generosity overflows and causes havoc, as in 1891. Such misadventures did not then affect the resources (fishing, agriculture, husbandry), only the infrastructure. It was the development of human economic and industrial activities and their integration into international trade that gradually 'diabolized' the 'infant Jesus', since the decrease in anchovy biomass had no negative effect so long as its exploitation remained below its reproductive potential.

ALFONSO PEZET – OR EL NIÑO OF THE SECOND KIND

Alfonso Pezet, on behalf of the Lima Geographical Society, took up his colleagues' data to present them in London in 1895 in a talk on 'The Countercurrent El Niño on the Coast of Northern Peru'. This was a historic date; firstly, because it marks the scientific recognition of the El Niño current, and secondly, because it portrays its variability with the appearance of exceptional floods, as that in 1891. Pezet writes:

Whereas nearly every summer here and there, there is a trace of the current along the coast, in that year it was so visible, and its effects were so palpable by the fact that large dead alligators and trunks of trees were borne to Pascamayo from the north, and that the whole temperature of that portion of Peru suffered such a change owing to the hot current which bathed the coast.

Finally, he linked El Niño to the climate, again writing: 'that this hot current has caused great rainfall in the rainless regions of Peru appears as a fact …'. Later studies showed that there is no direct relation between the rains and the warm current, but that they both belong to the same phenomenon at the level of the equatorial Pacific. Nevertheless, Pezet's article already posed the question of the relation between the ocean and the atmosphere in the climate system; the answer was given seventy years later by Jacob Bjerknes.

With the exceptional characteristics of El Niño 1891 in mind, Victor Eguiguren went through the archives from Spanish missions to look for similar rainy events. He thus went back to 1578, a particularly catastrophic year, then noted ten years during which there were very heavy rains within the hundred years preceding 1891: 1790, 1804, 1814, 1828, 1845, 1864, 1871, 1877, 1878 and 1884. According to him, these rainy years were due to the unusual presence of the warm water of the El Niño current, as in 1891; and he was right! So much so that the term El Niño became used exclusively for exceptional events, robbing the fishermen of Paita of their special relationship with a rather good-natured God. Therefore, thanks to the 1891 event, we owe the advent of El Niño of the second kind (seen as an exceptional and threatening economic phenomenon) to the honourable members of the Lima Geographical Society.

EL NIÑO OF PLENTY

If, since then, researchers have dug into the past, it has been with the idea of predicting future El Niños. The American biologist Robert Murphy was a witness to the '1925 vintage', which he considered to be the most powerful since 1891 and which he described as *'El Niño de abundancia'*. He recalled the writings of the witnesses of the 1891 event, which are quite the opposite of today's apocalyptic presentation:

> If the sea was full of wonders the land was even more so. First of all the desert became a garden … The soil is soaked by the heavy downpour, and, within a few weeks, the whole country is covered with abundant pasture. The natural increase of flocks is practically doubled, and cotton can be grown in places where in other years vegetation seemed impossible.

It was the exploitation of guano and then marine resources that would tarnish this reputation, making of El Niño an economic catastrophe; sometimes rightly so, but sometimes mistakenly, especially as regards the fisheries for which it serves as a scapegoat to mask the effects of overfishing.

THE EXPLOITATION OF GUANO

Guano is a manure produced from the droppings of myriads of sea birds that concentrate on the islands off Peru. Such a proliferation of these birds is due to an abundance of their prey, anchovies, which in turn feed on the plankton that is abundant in these fertile waters. In fact, just as on land, certain regions of the sea are deserts, while other parts receive masses of nutrients, offer optimum conditions for plankton growth and constitute true marine 'meadows' where fish can find food in abundance. These fertile ocean regions are those where deep cold water rich in nutrients comes to the surface. This is what happens in the Pacific along the coasts of Peru and California and, in the Atlantic, off the western coast of Africa (Mauritania/Senegal, in the north; Namibia/South Africa, in the south). Along these coasts, the trade winds push the surface water towards the open sea, creating a 'vacuum' in the coastal sea which is filled by the advection of deeper water towards the sea surface. This phenomenon is referred to as upwelling (see Chapter 3, Fig. 3.6). In Peru, anchovies are a real feast for millions of sea birds that nest on the islands where they produce the precious guano. Exported to other parts of the world, it was an important source of foreign currency for Peru.

Guano exploitation intensified towards the end of the 19th century and in 1909 the Peruvian Government created a state company to control the exploitation. Guano became a very important resource. The guano suppliers became a pressure group for which El Niño became the enemy: first of all because the heavy rains washed away the birds' droppings thus reducing the resource; then the warm-water invasion forced the anchovies to migrate towards the south or to deeper waters to reach a more favourable habitat. Thus the birds were condemned, at best, to a severe diet, or, at worst, to a fatal famine; they are the major victims of El Niño! The 1957–58 episode halved their population; 15 million birds disappeared. Moreover, this gave ideas to newcomers on the scene – industrial fishermen.

INDUSTRIAL FISHING AND FISH MEAL

After the war, two factors put a stop to the action of the guano pressure group that was blocking the development of industrial fishing which would have reduced the guano-producing birds' food source. The first factor was the competition from Chilean nitrate: the guano production decreased, so did foreign currency income, and, naturally, the pressure group's influence. After that it was the greed of businessmen; for them to make a profit, poultry and American cattle were forced to develop a taste for fish meal. However, the production of feed from California sardines collapsed as a result of over-exploitation during the war when the fishing quotas were no longer respected. In his novel *Sweet Thursday*, John Steinbeck humorously writes: 'The canneries themselves fought the war by getting the limit taken off fish and catching them all. It was done for patriotic reasons, but that didn't bring the fish back. As with the oysters in *Alice* [*in Wonderland*], "They'd eaten every one".' Attention was then focused on Peru. The Peruvian Government lifted the protection measures imposed by the guano pressure group, and thus industrialized fishing grew rapidly: fish catches increased from less than 100,000 tons in the early 1950s to more than 10 million tons in 1970–71. To avoid 'unfair' competition, in the 1960s, some people even proposed that the birds should be culled in order to maintain their populations at the lowest possible level that would still be compatible with biodiversity conservation, and to increase the fish-meal production. And yet, in 1970, this output already represented more than a third of the world's fish-meal production. It was then (in 1972) that El Niño appeared and, in 1973, catches fell to 1.5 million tons, to the delight of soya producers who then took over from the fish-meal market. No one was interested in who was benefiting from the crime, but only with designating the culprit: El Niño, the scapegoat for the industrial fisheries which, from 1962 to 1971, caught on average 9 million tons per year from a biomass estimated at 20 million tons.

El Niño's impact is twofold. The warm water that it moves overlies the cooler fertile waters, the natural habitat of anchovies, which are then forced to migrate and thus are not accessible to the fishing gear. The catches decrease sharply (during the 1982–83 event, they were below 100,000 tons), which protects the resource. But, on the other hand, the conditions necessary for reproduction of the fish stocks are rather poor: the larvae and juveniles have great difficulty in surviving because of the decrease in primary productivity linked to the invasion of the less-fertile warm waters. Nevertheless, after the historical El Niño of 1982–83 and the more moderate one of 1986–87, the stock was replenished and the catches increased in size, exceeding 6 million tons between 1992 and 1996 and reaching 10 million in 1994, superbly indifferent to the 1992–94 El Niño. This indicates that fishing is regulated not only by El Niño but also by another type of climatic variability on a decadal scale and, obviously, by the fishing effort. The consequences of the 1997–98 event are particularly interesting (see Chapter 7).

SOUTHERN OSCILLATION AND THE INDIAN MONSOON

In parallel with the discovery of El Niño, which is considered to be a local marine phenomenon, there is another atmospheric phenomenon: the 'southern oscillation'.

We owe a lot of our knowledge on the subject to the curiosity of Her Britannic Majesty's subjects. In fact, they developed observatories, including meteorological ones, all over the world. The story starts in 1877, when a weak monsoon caused a severe drought in India, leading to the deaths of tens of thousands of people as a result of famine. That same year, the *seca* (= dry, in Portuguese) period affected north-eastern Brazil causing the deaths of 500,000 people of whom 100,000 lived in Fortaleza. The previous year, the drought that affected New Caledonia was considered to be one of the causes of the great Kanaka revolution.

Meteorology was then a young science and Henry Blanford, 'the first imperial meteorological reporter', appointed in India, noted that this event corresponds to an abnormal rise in atmospheric pressure. At the risk of ruining the suspense, let us recall that 1877 was a particularly rainy year for Peru and was recorded as an El Niño year by Eguiguren. Having recourse to the observatories in the Indo-Pacific zone (Island of Mauritius, Australia, New Zealand), Blanford showed that the abnormal atmospheric pressure recorded in India was spread over the whole region that the drought affected, including Australia. He and his successors tried to link this discovery to other precursors of the monsoon season. They attempted, without much success, to relate it to sun spots and snowfall in the Himalayas in the months preceding the monsoon.

It was Gilbert Walker, director of the observatories in India from 1904 to 1924, who took a decisive step forwards, thanks to his mathematical knowledge. He did not have a computer, but instead, an abundance of workers whom he turned into 'human calculators'. He systematically looked for correlations between the monsoon and meteorological observations around the world. Hence, in 1909, he established the first equation for monsoon prediction:

rainfall of the Indian monsoon = –0.20 (precipitation in the Himalayas) –0.29 (pressure in Mauritius) +0.28 (average pressure in South America) –0.12 (rainfall in Zanzibar).

He did the same to predict flooding of the Nile, as well as precipitation in Australia. By synthesizing these statistical relationships, he demonstrated three coherent oscillations in the atmospheric parameters between the major regions of the Earth's surface. He defined the southern oscillation as a 'see-saw' movement between pressure zones and areas of precipitation in the Indo-Pacific region (from Egypt to Australia), and the pressure over the Pacific region. An increase in pressure and a reduction in precipitation in the Indian Ocean corresponded to a reduction in the pressure in the Pacific and vice versa. For South America, Walker had data for Chile, Argentina, Brazil and Paraguay. If he had had information for Peru, he would most certainly have linked the southern oscillation and El Niño. He also defined the same type of oscillations in the North Atlantic region, between the Azores and Iceland (North Atlantic Oscillation), and in the North Pacific region between Hawaii and Alaska (North Pacific Oscillation). Based on statistical relationships, he established a Southern Oscillation Index, which defines the see-sawing of the atmosphere between the Indian and Pacific Oceans, to serve as a tool for the prediction of monsoons. Nowadays, a much simpler index is used: the atmospheric pressure difference between Tahiti and Darwin (southern oscillation index, see Chapter 4, Fig. 4.1). Walker, who thus laid the foundation for long-term meteorological predictions, did not convince his fellow citizens. The predictive capabilities of these relationships were low; even the relationships themselves were somewhat doubtful. In fact, they were purely statistical; there was no physical mechanism or hypothesis underlying them.

The loss of interest in the southern oscillation lasted until 1957–58, the International Geophysical Year. Nonetheless, the subject might have come up again in 1933 if John Leighly's work had been given more attention. He associated the pressure differences between the two extremities of the equatorial Pacific to oceanic and meteorological conditions of the tropical Pacific. He pointed out that the more the atmospheric pressure gradient between the east and west is pronounced in the central Pacific, the more strongly the trade winds blow, the more the sea temperatures are low, and rainfall less abundant, and vice versa. He was thus describing ENSO. Walker's interest was in the Indian Ocean monsoon and Leighly's in the climate of the Marquesas Islands; they were both talking of the same mechanism, the southern oscillation, but no one made the connection.

THE INTERNATIONAL GEOPHYSICAL YEAR: 1957–58

Leighly's intuition was all the more surprising, given that he had very few data on ocean-surface temperature. For that is where the problem lies: the ocean, which sailors know well, remains for scientists a *terra incognita*, so to speak. The exploration of the atmosphere was carried out as a matter of course; we live in it and we measure directly its caprices which affect all our activities – farming, commerce, leisure, etc.

Meteorology became a science in the 17th century, with the development of instruments to measure temperature, pressure, humidity, wind speed and direction, and with the establishment of weather observatories. We owe the construction of the first meteorological network to Ferdinand II Medici. From 1653 onwards he financed the construction of thermometers, barometers and hygrometers for the benefit of renowned scientists in eleven cities of Europe. This network functioned until the Vatican, taking exception to the initiatives of the Grand Duke, forced him, in 1667, to close his Accademia del Cimento, founded ten years earlier to study natural phenomena. His meteorological observation network did not survive this blow.

Meteorology remained for a long time a minor branch of physics. The Royal Observatory of Paris, founded in 1670, was more concerned with the nobility of the stars' motions than with the capricious behaviour of the unpredictable sublunar world. For no apparent reason, meteorological observations were interrupted from 1754 to 1785. To use present-day terminology, physicists would say that meteorology was a 'soft' science compared to astronomy, which, thanks to Isaac Newton and Simon Laplace, had been treated mathematically, thus allowing planetary trajectories to be calculated with precision. Physics brought order to the universe, whereas meteorology, a priori, gave an image of disorder, to such an extent that Auguste Comte excluded it from his classification of the sciences – for bad behaviour, one might say!

Even so, a hundred years later, the science of non-linear system dynamics or 'chaotic systems' popularized by Lorenz and his famous butterfly effect was born out of meteorology. It is therefore not surprising that it was a naturalist, Jean-Baptiste de Monet, a knight of Lamarck, and not a physicist, who published a *Meteorological Directory* from 1799 onwards, and who, in 1807, proposed unsuccessfully the establishment of a Central Meteorological Bureau.

The trigger came in 1850 with the creation of scientific meteorological societies, the transmission of observations via electric telegraphic Morse code, and 'thanks to' the destruction of the allied fleet besieging Sebastopol during the Crimean War, by a storm on 14 November 1854. According to Urbain Le Verrier, director of the Paris Observatory, this disorderly retreat could have been avoided if there had been an international network of meteorological observations. As from 1858, the Paris Observatory published a daily *Bulletin météorologique international*, and each country set up its own meteorological services. International co-operation was organized (international congresses in Leipzig in 1872, in Vienna in 1873 and in Rome in 1879) to culminate, in 1879, in the creation of an International Meteorological Committee, the precursor of the present World Meteorological Organization. Thus, in 1877, at the start of the history of the southern oscillation, meteorological science and observation networks were well established.

At that time, nothing equivalent existed for the ocean, which was not considered a proper subject of research. A circular sent to scientists by the founders of the Meteorological Society of France on 17 August 1852 indicates the thinking at that time:

> Of the three branches that constitute the entire field of geophysics, geography and geology are the only ones hitherto to have led, in France, to the creation of a centre where all the facts and information arising from the study of these sciences can be assembled and later, through extensive publicity, widely distributed. Placed between them and acting as a natural link, meteorology still lacks this powerful mode of action and progress.

As for the ocean, which covers 70% of the planet, it had no place in geophysics! During the 19th century, it simply represented a basis of power for conquest and trade.

The first oceanographic campaigns, like that of the British HMS *Challenger*, which sailed the seas from 1873 to 1876, responded to a maritime and colonial vocation rather than to scientific concern, even if, at the end of the day, the scientific harvest was fruitful. Germany and the Netherlands also organized such campaigns in order to help the exploration of their colonies. Until the Second World War, oceanographic research remained national: competition overrode international co-operation. It was only in 1960 that the Intergovernmental Oceanographic Commission was created by UNESCO: eighty years after the first International Meteorological Committee! These eighty years give us an idea of the gap that had existed until then between knowledge of the atmosphere – and ignorance of the oceans. It was the International Geophysical Year of 1957–58 that made physical oceanography a recognized branch of geophysics. This exploration of the planet was organized to coincide with maximum solar activity, which is a source of several different magnetic phenomena. Sixty-seven countries, covering all the earth sciences, came together and, for the first time, the oceans were the object of simultaneous and co-ordinated systematic observations. The equatorial zone of the Pacific, from the Galapagos to New Guinea, was covered; generously, El Niño manifested itself with a violence unknown since 1941. The scientific community thus discovered that the invasion of warm water relates to the entire equatorial zone right up to the 180° meridian; El Niño changed scale! The researchers also discovered that the trade winds were weak and that rainfall was abundant in the central equatorial Pacific, hence confirming Leighly's forgotten results. The question: Is there a cause-and-effect relationship between the ocean, with its warm anomalies (El Niño), and the atmosphere, with its Pacific-wide perturbations (southern oscillation)? could then no longer be evaded.

Let us recapitulate. The southern oscillation shows clearly the see-sawing of the atmosphere in the Indo-Pacific region, which we can summarize as follows: the higher the pressure rises in the Pacific, the more it decreases in the Indian Ocean and vice versa. The Indian monsoon is affected by this oscillation, since high pressures in the Indian Ocean correspond to a failure in the monsoon; such failure can have extremely harsh consequences. Measurements taken during the International Geophysical Year of 1957–58 showed that this programme, the IGY, was carried out during a time of low pressure in the Pacific, and thus of high pressure in the Indian Ocean. The low value of the southern oscillation index corresponded to an El Niño episode, which was not confined to the coasts of South America. From these results, confirmed by the El Niño event of 1965–66, Bjerknes proposed a simple hypothesis of the interaction between the ocean and the atmosphere linking these two phenomena – El Niño and the southern oscillation. To understand this interaction fully, one needs to have a general outline of the oceanic and atmospheric circulations in the tropical region.

3 Ocean–atmosphere coupling

THE EARTH'S ROTATION AND THE CORIOLIS FORCE

The two main actors of the climate drama, the ocean and the atmosphere, are two fluids whose dynamics are governed by the same factors: gravity and pressure. If the Earth did not rotate on its own axis, the winds would blow directly from the high polar pressures to the low equatorial pressures (it would be the same for marine currents), whereas the warm air would rise above the equator and return to the poles at the top of the troposphere, thus exporting part of the excess heat.

We are not aware of this rotation, despite the fact that it is dragging us along at a speed of 1,700 kilometres per hour. It is more natural to imagine that it is the stars that are moving around an immobile Earth. The Polish astronomer, Nicolaus Copernicus, in 1542, published his concept of a system in which the Earth was not the centre of the universe but simply a satellite of the Sun, revolving around it like a top. It took 300 years to show experimentally the rotation of the Earth on its own axis. In 1851, the French physicist Leon Foucault suspended a 28-kilogram pendulum from a 67-metre wire, from the dome of the Pantheon, in Paris. The pendulum was equipped with a stylet that traced its trajectory in sand at each return of the pendulum; it could be seen that the plane of the pendulum accomplished one complete turn in a clockwise direction in thirty-two hours. This experiment, using a much smaller pendulum, can be seen today at the Conservatoire National des Arts et Métiers de Paris (Paris National Conservatory of Arts and Crafts).

The rotation of the Earth was hence shown, as well as its effects on the movement of bodies: here, the pendulum's plane of oscillation seems to rotate relative to the Earth, whereas, in reality, it is the Earth that revolves around the plane of the pendulum. In mechanics, a movement is generally associated with the force that caused it: the force of gravity for an apple that falls from a tree or for satellites revolving around the Earth. By analogy, a force has been 'invented' to translate the effect of terrestrial rotation on the displacement of bodies. This is the force of Coriolis, named after the French mathematician Gaspard Coriolis who gave the corresponding mathematical explanation in 1836. This force, which applies to all bodies in movement on a solid object in rotation, deviates the moving body to the right in the northern hemisphere, and to the left in the southern hemisphere. This force is at its maximum at the poles and nil at the equator, and is generally negligible on the surface of the Earth compared to other forces. Even if on the highway you are subject to it without being aware of it, it is not the force that should be blamed in case of an accident, nor if you miss your target while shooting with your rifle! However, this force becomes important on long trajectories and slow continuous movements such as those of atmospheric and marine currents. Three examples serve to better illustrate the Coriolis effect.

At the start of the twentieth century, the Norwegian Fridtjof Nansen noticed, during his extensive travels through the Arctic ice, that the ice floe did not drift in the same direction as the wind, but to the right; the Swede Walfrid Ekman proposed a theoretical basis for this observation by showing that a surface current flows to the right of the wind direction in the northern hemisphere, and to the left of it in the southern hemisphere.

Now let us imagine releasing an object from the top of the Eiffel Tower. It does not reach the ground at a point vertically below where it was dropped; it lands some 10 centimetres eastwards. Why? Because the top of the tower, in twenty-four hours, traces a larger circle than that traced by its base; the top therefore travels much faster. The object released at 'zero initial speed' has thus, in reality, an eastward velocity relative to the ground. During its 8-second fall, the object continues moving in an easterly direction with respect to the ground.

The firing of a missile from the equator towards the North Pole is the last example. When the projectile leaves the launcher, an eastward deplacement is superimposed on its firing speed towards the north. But, although the speed of the Earth's surface eastwards is at a maximum at the equator, the missile appears to follow a straight line, since it is going at the same speed. While moving to the north, it maintains its original speed, whereas the part of the Earth's surface under it is rotating at an ever decreasing speed. With respect to the Earth, the missile is moving not only towards the north but also eastwards at an ever increasing speed.

The general characteristics of the atmospheric and oceanic circulations are a result of the equilibrium between the pressure forces and the Coriolis force. This 'geostrophic' equilibrium implies that these two forces are, at one particular moment, of equal intensity and opposite directions. The pressure force is always from a high to a low pressure, whereas the Coriolis force is perpendicular to the speed. At equilibrium, the speed is necessarily perpendicular to the pressure gradient (tangential to isobars) and not in the direction of pressure variations, as would be the case if the Earth did not rotate (Fig. 3.1). Hence, in the northern hemisphere, the wind rotates in a clockwise fashion around anticyclones and anticlockwise around depressions. The opposite happens in the southern hemisphere; and the same goes for sea currents.

THE METEOROLOGICAL EQUATOR AND MERIDIAN CIRCULATION: HADLEY CELLS

On either side of the equator there are always anticyclones or zones of high atmospheric pressure. Their position and their intensity vary with the seasons: they are stronger and move to higher latitudes during the summer season of the hemisphere they are in. For example, there is the Azores anticyclone over the North Atlantic which is responsible for the fine summer weather over western Europe, and the Saint Helena anticyclone in the South Atlantic. Equivalent structures exist over the Pacific, bringing high pressures to northern California and to Easter Island in the South Pacific.

The winds blow around these anticyclones; their equatorial branch constitutes the westward trade winds, the constancy of which was much appreciated by navigators during their voyages around the world. The northern and southern trade winds converge along a line called the intertropical convergence zone (ITCZ) or 'meteorological equator'

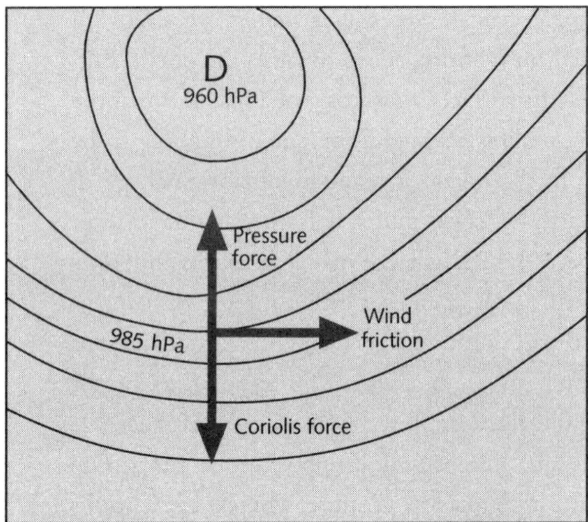

Fig. 3.1
Pressure field, Coriolis force and winds in the northern hemisphere.
In the absence of the Earth's rotation, the winds would blow from areas of high pressure towards areas of low pressure (D). But the Coriolis force, owing to the rotation of the Earth, causes the winds to deviate to the right in the northern hemisphere, as in this figure, and to the left in the southern hemisphere. The pressure force balances the Coriolis force (geostrophic equilibrium) and the wind is tangent to the isobars. In the northern hemisphere, the winds turn in the direction of the hands of a clock around areas of high pressure, and in the opposite direction around areas of low pressure. It is quite the opposite in the southern hemisphere.

(Fig. 3.2): this zone constitutes the famous 'doldrums' which sailors used to dread so much because they could be becalmed for several weeks; and even the pioneers of transoceanic aviation encountered strong turbulence that was dangerous for their fragile aircraft. All along the ITCZ, the meeting of trade winds, heavily loaded with oceanic humidity, gives rise to ascending masses of air that generate turbulence and heavy rainfall when the water vapour condenses as the rising air expands with increasing altitude. The trade winds transform their horizontal kinetic energy into vertical kinetic energy, so that, at sea level, the winds are weak under the regions where the air is ascending, greatly to the detriment of sailors.

At altitude, this rising flow of air diverges towards the north and the south and, becoming cooler and drier, it descends again in the subtropics

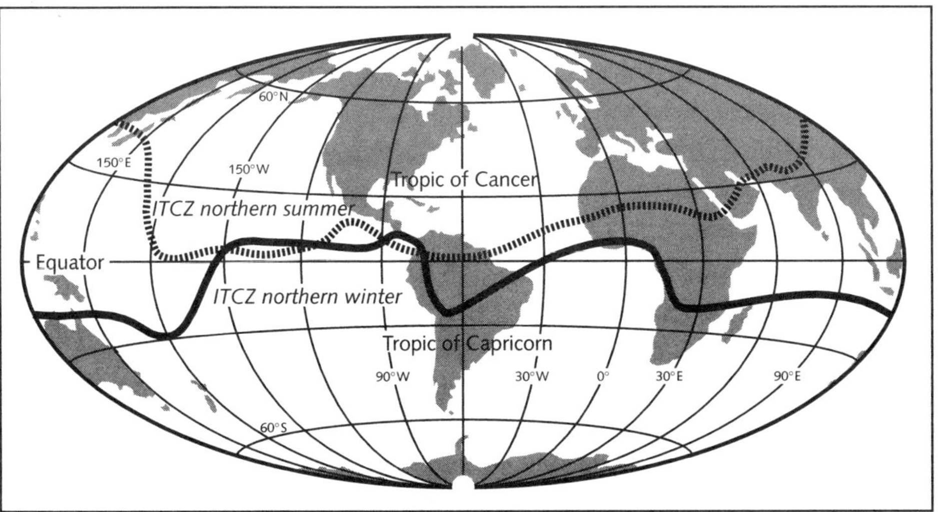

Fig. 3.2
Position of the intertropical convergence zone (ITCZ) or 'meteorological equator' during the two seasons.
The north-east trade winds in the northern hemisphere, and the south-east trade winds in the southern hemisphere, deviated by the Coriolis force, all blow westwards. They blow towards a calm region, the meteorological equator, known as the doldrums in the days of sailing ships, because of the overcast weather and lack of winds there.

at approximately 30° latitude, causing anticyclonic situations. We give the name 'Hadley cell' (Fig. 3.3) to the latitudinal loop that becomes established between the meteorological equator or 'doldrums' – a zone of low pressure – and the heart of the anticyclones, in the northern as well as the southern hemisphere. Although somewhat disturbed by the monsoons, this pattern of circulation is also found over the continents all around the Earth. The regions of descending air correspond to dry climates that are the cause of the great deserts in the two hemispheres, such as the Sahara in the north and the Kalahari in the south.

Navigators quickly learned to take advantage of these various wind patterns. At the head of a Portuguese fleet, Vasco da Gama, in search of the route to India between 1497 and 1499, already had a thorough knowledge of the wind field in the Atlantic. Rather than becoming becalmed in the doldrums off the coast of Africa, he took

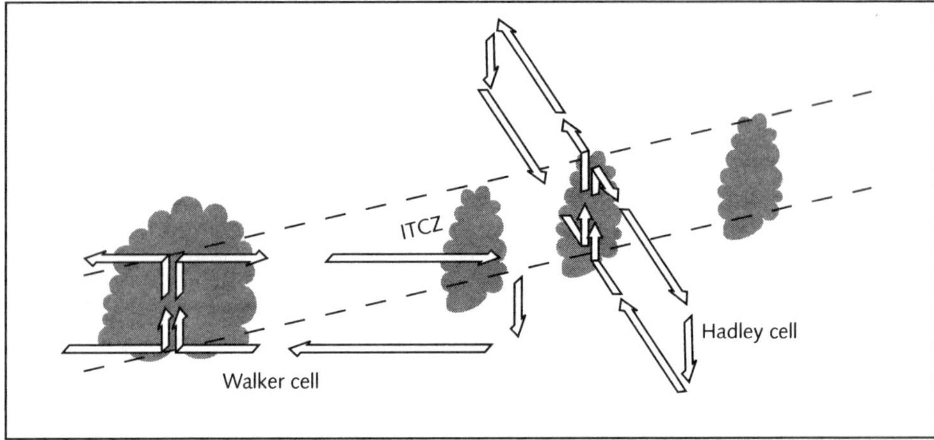

Fig. 3.3
Intertropical atmospheric circulation of Hadley and of Walker.
In the subtropical region, the atmospheric circulation on a large scale can be broken down into a meridional circulation (north–south and south–north), with two Hadley cells (one to the north, the other to the south of the equator) and a longitudinal circulation (east–west and west–east), with a Walker cell over each ocean. This is a simple way of presenting the three-dimensional Hadley–Walker circulation. It is characterized by zones of uplift, hence of rainfall, near the equator and on the western side of the ocean (Indonesia, Amazonia) and zones of descending dry air in the tropics on the eastern side of the ocean and on the neighbouring continents, marked by the ring of major deserts: in the northern hemisphere, the deserts of Mexico–Texas–Arizona, the Sahara and the Gobi (of complex origin because of the proximity of the Himalayas), and, in the southern hemisphere, the deserts of the Andean highlands, the Kalahari and the Australian desert.

a bearing from the Cape Verde Islands, towards Brazil, taking advantage of the north-east trade winds. He then sailed southwards off South America, turning for the Cape of Good Hope when he had crossed latitude 30°S, so as to be able to take advantage of the westerly winds. A few decades later, the Spanish did likewise in the Pacific. Sailing from Mexico in 1527, Alvaro de Saavedra filled his sails with the north-easterly trade winds right across to the Philippines, thus opening the route for Spanish galleons from Acapulco to Manila. The outward journey was made between 5° and 10°N, whereas the return journey, reconnoitred by Andrés de Urdaneta in 1565, was 30° farther north so as to take advantage of the dominant westerly winds on the northern side of the anticyclone. Navigation at lower latitudes was therefore governed by two requirements: avoiding the doldrums and making the most of the fair winds, either the westerlies between 30° and 40°N, or the trade winds in the tropics.

The outcome of the major transoceanic sailing races, in which the vessels sail faster against the wind, depends very much on the choice of the best possible meteorological track, as is shown by the advice provided by 'track' meteorologists.

The position of the ITCZ or meteorological equator does not coincide with the geographical equator; it changes with the seasons, being always 'drawn' towards the hemispheric summer. Even so, it is nearly always situated north of the equator, even during the southern-hemisphere summer, because of the dissymmetry in the distribution of the oceans and the continents between the two hemispheres.

A BIT OF COMPLEXITY: THE WALKER CIRCULATION

The atmospheric circulation is, in reality, more complex. The trade winds that blow from east to west displace the hot surface water in the same direction; this displacement is compensated in the east by the upwelling of colder water. Temperature differences thus arise between the two sides of the tropical ocean and, of course, between the continents and the oceans. This is particularly true in the western Pacific where there is a vast reservoir of sea water whose temperature exceeds 28–29°C. It is the part of the world where the ocean transfers the most energy to the atmosphere; convection is very intense there. The air, heated and loaded with humidity through contact with the ocean, rises; during this process, the water vapour condenses giving rise to well-developed cumulo-nimbus cloud formations carrying the rain that falls generously on Indonesia. This convection is the ascending branch (low atmospheric pressure) of an atmospheric circulation cell along the equator (Fig. 3.3). The descending branch of this cell, linked to the ascending branch by a west-to-east wind at altitude, is situated to the east over colder ocean water and corresponds to high atmospheric pressures and to a supply of dry air; rainfall is, in fact, very rare on the coasts of Peru and northern Chile. The trade winds, which blow from east to west at the ocean surface, complete this circulation cell to which Bjerknes gave the name Walker in honour of the discoverer of the southern oscillation. There is in fact a direct relation between this circulation and the southern

The pedagogical value of cyclones

A cyclone is a spectacular meteorological phenomenon not only in itself but also in its consequences when it hits the coast. Cyclones often come to mind during El Niño/La Niña episodes, and serve as an introduction to the notions of ocean–atmosphere exchange and of the transport of heat. Ignorance of these processes hinders understanding of ENSO.

Cyclones are meteorological perturbations in the tropics and are given a wide variety of names in the various regions in which they occur. In the eastern Pacific, in Mexico, in California and in the western Atlantic, the name *hurricane* is applied, after the Mayan wind god Hurraken. In the Indian Ocean, the term *cyclone* is used, from the Greek term meaning 'to wind like a serpent'; it is this term that is generally used in French. Off the northern coast of Australia, it is called *willy willy,* and in the north-west Pacific, the term *typhoon* is preferred, although its etymology is disputed: from the Portuguese *tufão,* from the Arabic *tufan* meaning 'eddy', or from the Chinese *tai-fung,* meaning 'strong wind'?

As a veritable safety valve, or a kind of energy pump, it allows the ocean to offload to the atmosphere and towards the temperate regions its excess heat accumulated in the tropics. Data on cyclone tracks in the North Atlantic Ocean show that they are created around 10°N latitude and break up between 30° and 40°N, some of them ending up as storms on the coasts of western Europe.

A cyclone cannot be formed unless the temperature exceeds 27°C in the upper 50 metres of the sea surface. Such a temperature allows intense evaporation and the transfer of humidity from the ocean to the atmosphere to occur. Each year, several hundred cyclones are created over the tropical ocean, cyclone activity being greatest when the surface-water temperature reaches 28–29°C. This is a necessary but not sufficient condition for the formation of tropical storms and cyclones. The existence of strong winds at altitude blocks the formation of cyclones; they also explain the absence of cyclone activity in the Atlantic when there is an El Niño episode in the Pacific Ocean. Also, cyclones can only form at a certain distance from the equator at latitudes where the Coriolis force is strong enough to cause whirlwinds. About 10% of the whirlwinds thus created evolve into cyclones.

This transformation of heat energy into kinetic energy represents an extreme case of ocean–atmosphere exchange, and underscores a fundamental fact: under the atmosphere in which we live – the troposphere – with a thickness of some 15 kilometres, transparent to the solar radiation, it is the land and the oceans that absorb most of the solar energy; they constitute the 'radiator' of the atmosphere which is thus heated from below. A cyclone underscores – and how! – the dynamic character of the atmosphere. The air flows in from everywhere towards the centre of low pressure; this flow continues towards the eye of the cyclone and can only escape upwards, forming huge cumulo-nimbus clouds.

In a cyclone, this liberation of latent heat increases instability and strong upward movement of air. The cyclone takes up more energy so long as it remains in contact with the hot surface water, since heat and humidity are vital to its survival. When the hot water of the Pacific moves eastwards, the cyclones follow this displacement; they do not occur in the western part of the ocean. Their frequency is highest in the central Pacific Ocean, in the area bounded by Polynesia, Hawaii, and the Cook Islands, and some are formed in the east, striking Central America which is generally more concerned by the cyclones of Atlantic Ocean origin. Hurricane Pauline, which hit Acapulco and Oaxaca, in Mexico, in the autumn of 1997, the most violent for thirty-five years, was formed in the Pacific.

When a cyclone reaches the coast or ocean areas that are not sufficiently hot, the energy from latent heat becomes less than the kinetic energy dissipated and the cyclone breaks up; the sea-surface temperature is in fact a critical parameter of the cyclone's heat supply.

oscillation. The intensity of the trade winds is proportional to the difference in atmospheric pressure between the eastern and western Pacific; so much so, that the index that specifies this difference also constitutes a measure of the intensity of the Walker circulation. A high value of the index corresponds to intense trade winds, and vice versa.

An analogous cell, though of smaller size, can be observed over the Atlantic. The situation is more complex in the Indian Ocean, which can be described as a semi-ocean, since its northern side extends only to 25°N, being limited by the huge continent of Asia. There is thus established, all along the equatorial belt, a series of cells in which there are alternating zones of upward convection (low atmospheric pressures, substantial heat transfer to the atmosphere, abundant rainfall), as in the western Pacific, equatorial Africa and Amazonia, and of downward air flow (high pressure and much drier air).

In reality, these two types of cell are not independent; the air is simultaneously affected by the Hadley and the Walker circulations which represent the meridional and zonal decomposition of the air movements, as is done in physics with the parallelogram of forces. Thus, during an El Niño episode, which corresponds to a weakening of the Walker circulation, the spreading of the hot water towards the central and eastern Pacific increases the heat exchange with the atmosphere and intensifies the Hadley cell, hence the heat transfer towards high latitudes. Thus starts the process that explains why the consequences of El Niño are not limited to the tropics even if they are easiest to identify there.

THE OCEANIC THERMOCLINE

Since the ocean gets its energy 'from overhead', the temperature decreases from the surface to the bottom, with markedly different vertical gradients from one region to another. In the polar regions, the low surface temperatures, even in summer, do not allow a marked thermal gradient to become established. Quite the opposite is true at low latitudes. In the tropics, the temperature, which is very high at the surface (25–30°C), decreases sharply at a certain depth: this is called the thermocline (Fig. 3.4). Below it, the gradient is again very low, with a temperature below 5°C in the intermediate and deep layers. In the temperate zone, a seasonal thermocline forms in summer.

The thermocline thus separates a warm surface layer from a cool deeper layer. On the oceanic scale, this warm layer has a much lower volume than that of the cool layer; overall, the ocean is cool, with a mean temperature of about 2°C. But the thermocline is also a strong vertical density gradient slowing down vertical water movements and diffusion. In the El Niño phenomenon we shall often refer to the thermocline, because it marks the base of the warm-water layer and because its depth variations provide an index of the phenomenon's evolution.

OCEANIC CIRCULATION: CURRENTS AND COUNTER-CURRENTS

The ocean is therefore not a homogeneous medium: at a given depth there are, as in the atmosphere, pressure differences that cause currents to flow to reduce these differences. The topography of the sea surface is a measure of the ocean pressure field: the higher the level of the sea surface, the greater the pressure. In the atmosphere a good approximation of the wind field can be gained from the observation of the pressure field. In the ocean, as in the atmosphere, the currents can be estimated from the sea-surface topography by applying the hypothesis of geostrophic equilibrium. Even so, sea-level differences are small and therefore hard to measure; for example, the Gulf Stream, one of the strongest ocean currents, has a slope of about 1 metre per 100 kilometres. It is remote sensing from space that has allowed us to overcome this difficulty. The first satellite measurements of sea level were made at the beginning of the 1970s. They have improved ever since and, today, the Franco-American satellite Topex/Poseidon, launched in 1992, covers nearly all parts of the world ocean and is capable of detecting differences of about 1 centimetre; that is, expressed in terms of pressure, 1 hectopascal. The wind, by generating sea-surface currents, causes differences in the pressure and the level of the ocean surface.

In the Pacific (the situation is similar in the Atlantic), the trade winds generate, on either side of the meteorological equator, two strong surface currents with a mean speed of 60 kilometres per day: the North Equatorial Current (between 10°N and 25°N) and the South Equatorial Current (between 2°N and 20°S) (Fig. 3.5). Both flow westwards, accumulating warm water on the western side of the ocean, creating a clear-cut rise in sea level in the warm-water pool of the western Pacific where the sea level is about 1 metre higher than on the eastern side. This gives rise to a series of currents to re-establish the pressure equilibrium. First there are the powerful western boundary currents, such as the Gulf Stream in the North

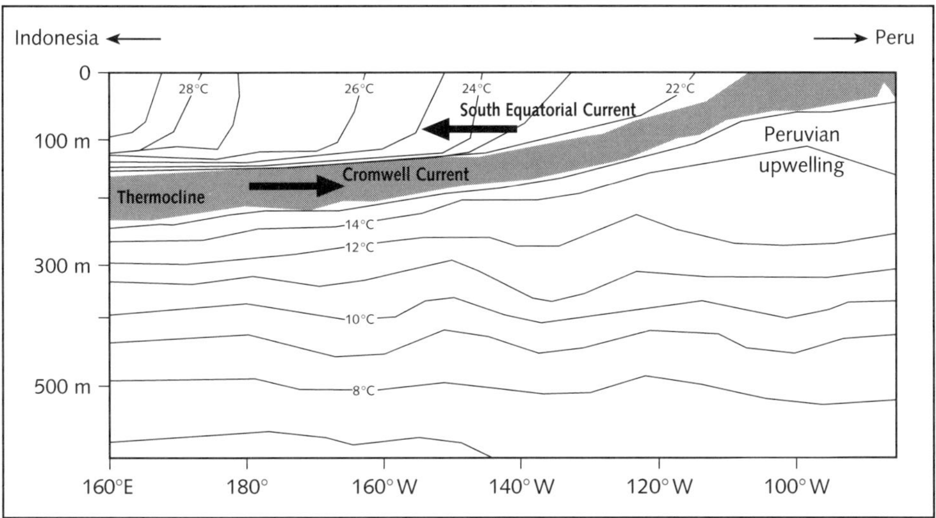

Fig. 3.4
Temperature profile of the equatorial Pacific showing the thermocline.
In a trade-wind regime, the temperature distribution is highly dissymmetric, with the upwelling of cool water off Peru, where the temperatures are below 22°C, and an accumulation of warm water towards Indonesia, where the temperatures exceed 28°C. The thermocline, which separates the warm surface water from the cool deeper water, is therefore deeper on the western side of the ocean basin. In the warm surface layer, the South Equatorial Current flows westwards. In the heart of the thermocline, the Cromwell Current flows eastwards. There is a circulation cell in the ocean symmetrical to the Walker cell in the atmosphere.

Atlantic and the Kuroshio in the North Pacific; they constitute the western sides of the major ocean gyres, corresponding to the anticyclonic circulation of the atmosphere. These gyres (in both hemispheres in each of the major oceans) are completed by the North and South Equatorial Currents, between which, along the meteorological equator, is the eastward-flowing Equatorial Counter-current which marks the oceanic equator. This east–west pressure gradient is the inverse of the atmospheric gradient; this sea-surface slope induces, along the geographical equator, a significant compensation current which flows beneath the South Equatorial Current within the thermocline, but in the opposite direction, at depths of between 50 and 200 metres in the west, and less than 100 metres in the east, where it promotes equatorial upwelling. This is the

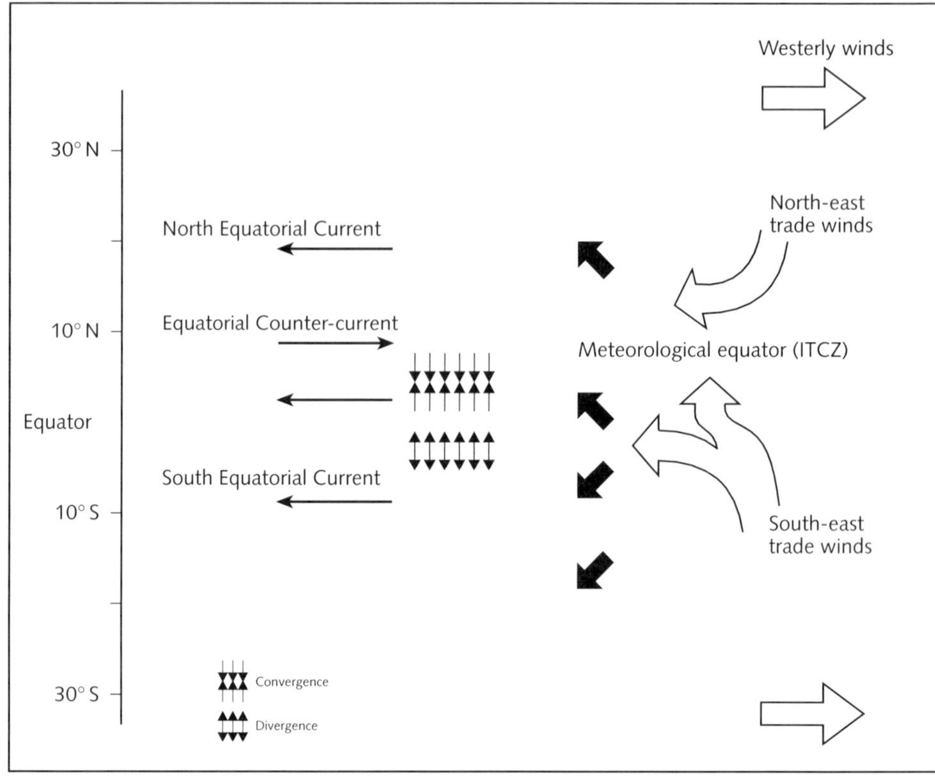

Fig. 3.5
Surface circulation in the tropical ocean.
The figure shows the relations amongst the winds, surface currents and water transport (black arrows) in the layer subject to the wind stress. The water flow is deviated to the right of the wind and of the surface current in the northern hemisphere, and to the left in the southern hemisphere, owing to the Earth's rotation. This causes divergences and convergences at the ocean surface.

Cromwell Current, named after the American oceanographer who discovered it in 1954. In the equatorial plan, therefore, the ocean has a circulation analogous to the Walker circulation in the atmosphere with symmetrical variations in the pressure field – a signature of the coupling between the two fluids.

TRADE WINDS, COASTAL UPWELLING AND EQUATORIAL DIVERGENCE

A map of the sea-surface temperatures of the equatorial Pacific (see Fig. D* on p. 58), shows, first of all, relatively low temperatures off Peru at latitude 8°S. During his voyage to the 'equinoctial regions of the New World', from 1799 to 1804, Alexander von Humboldt was the first to notice the contrast between the abnormally cool coastal water, in a region in which the air temperature was high, and the offshore water with a temperature exceeding 23°C. He attributed this anomaly to the transport of water from the Arctic by the current since named after him: the Humboldt Current. In reality, this current, which flows northwards along the coast of South America, has, because of the Coriolis force, a component to its left which carries the surface water offshore towards the open ocean. It is replaced, along the coast, by deep water originating at a depth of 200 to 300 metres; this water is cold and rich in nutrients (Fig. 3.6): this is coastal upwelling. The same phenomenon occurs on the coast of California and in Africa, on the coasts of Morocco and Mauritania in the north, and of Namibia in the south. All these regions thus fertilized by coastal upwelling are among the best fishing areas of the world.

The trade winds entrain the South Equatorial Current westwards, straddling the equator. With respect to the Coriolis force, this is a special situation, since the surface water is pushed to the right (northwards) to the north of the equator and to the left (southwards) to the south of the equator. So, at the equator, there is a divergence of the surface water and an upward 'draught' of deep water, as along the coasts of Peru. In the middle of this tongue of cold equatorial water, the temperature increases from 19°C to 20°C near the Galapagos Islands to more than 26°C at 180° longitude. The cooling of the sea surface by upwelling of deep water from just below the thermocline is made all the more difficult on the western side of the oceans by the fact that the hot surface layer brought by the South Equatorial Current gets thicker as we go westwards.

WAVES IN THE OCEAN

The term 'wave' is symbolic of the ocean. Even in this era of television, fibre optics and radio, the most easily observed waves are those that form on the surface of water when a stone is thrown into it; those that make the angler's float bob up and down. As most physical systems, the ocean and the atmosphere propagate their perturbations by means of waves.

* Figures A to L are to be found in the colour section on pages 57 to 64.

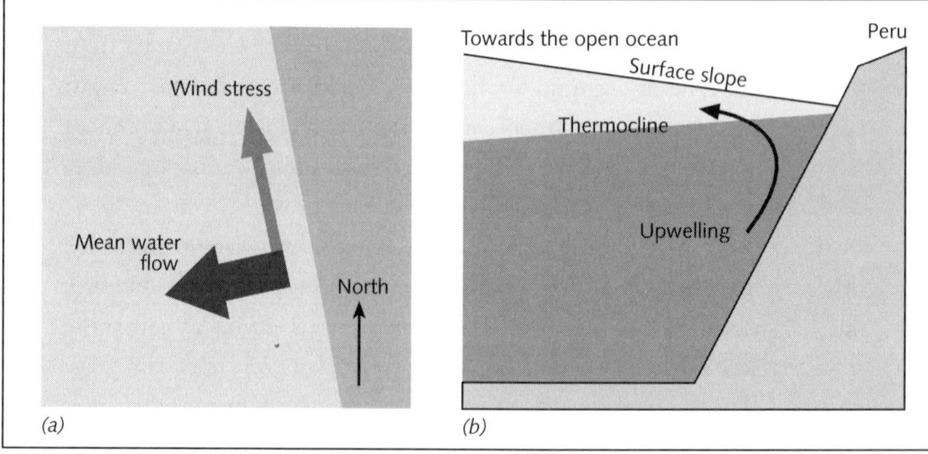

Fig. 3.6
Coastal upwelling.
(a) **By friction, the wind blowing parallel to the coast creates, in the southern hemisphere, a mean water transport to its left orthogonally to the direction in which it is blowing. Thus, on arriving in the trade-wind zone off the coasts of Ecuador and Peru, the Humboldt Current, which is flowing northwards, is deviated eastwards.**
(b) **In these conditions, the surface water is pushed offshore, thus creating a positive sea-surface slope starting at the coast. To compensate for this deficit, the underlying water near the coast, which is cooler and richer in nutrients, wells up towards the surface.**

When the ocean water is attracted by the Sun and the Moon, a tidal wave is created and propagates in the ocean; it becomes amplified on reaching the continental shelf and entering the coastal sea. Every wave is associated with movement of the particles of the fluid in which it is being transmitted. In the case of a sea wave, the movement is a circular one in the vertical plane without there being a displacement of mass. For waves with a long wavelength, horizontal movements involve a certain displacement of mass, so that they behave as real currents.

With the weak deviation of the trajectory at low latitudes, the change of direction of the Coriolis force at the equator and the presence of the thermocline at a few tens of metres beneath the sea surface, the equatorial ocean is an excellent 'wave guide' that facilitates the propagation of a perturbation or a vibration. This is the case of waves with a wavelength of several hundred kilometres. A stress applied to the sea surface by, for example, a variation in the wind around the equator generates two types of wave:

→ *equatorial Kelvin waves* (after the physicist Sir William Thomson, Lord Kelvin). These are planetary waves whose wavelength is long relative to the depth of the ocean. They propagate only eastwards, at a speed of about 200 kilometres per day. They are trapped at the equator by the Coriolis force which resists any tendency they might have to deviate. Let us take the example of a strengthening of the trade winds; this intensifies the divergence of the ocean water and promotes an uplifting of the thermocline. This perturbation triggers a Kelvin 'upwelling' wave which follows the equatorial wave guide, and the uplifting of the thermocline propagates eastwards. A Kelvin wave generated in the central Pacific reaches the coast of South America in two months. In contrast, a weakening of the trade winds or a period of westerly winds generates a Kelvin 'downwelling' wave, and the sinking of the thermocline propagates eastwards, hence the accumulation of hot water.

→ *Rossby waves* (described in 1939 by the meteorologist Carl-Gustav Rossby). These are planetary waves of long wavelength; they exist at all latitudes and propagate exclusively westwards. Their displacement speed depends on several factors, including wavelength, stratification of the ocean, the speed of the currents on which the Rossby waves are superimposed and, above all, the latitude. Such waves therefore propagate rapidly in the equatorial region, though at a speed two to three times less than that of Kelvin waves. They take about a year to cross the Pacific. Just as atmospheric disturbances generate Kelvin waves, they also produce, simultaneously, Rossby waves in the opposite direction, westwards; to a Kelvin *upwelling* wave there corresponds a Rossby *downwelling* wave, and vice versa.

When these waves reach the edge of an ocean basin, they are reflected by the coast, thus changing their direction of propagation. In the east the reflected Kelvin waves return westwards as Rossby waves; and, reciprocally, on the western side of the ocean, the Rossby waves return eastwards as Kelvin waves (see Fig. A on p. 57).

The vertical displacement of the thermocline when such waves are passing has a range of several tens of metres; this is reflected by a change in sea level of about 10 centimetres, which is easily detectable by satellite remote sensing (see Fig. B on p. 57).

To summarize: the atmosphere transmits to the ocean perturbations that propagate eastwards or westwards in the form of downwelling waves and upwelling waves that are reflected off the continental boundaries of the oceanic basins. In contrast to the atmospheric disturbances, which disappear in a few days, the oceanic response, through this interplay of waves, lasts several months. So it is the ocean that, after all, guides the perturbation. We can thus propose scenarios for the sequence of the various phases of El Niño by following the complex courses of these waves (see Chapter 4, Fig. 4.3).

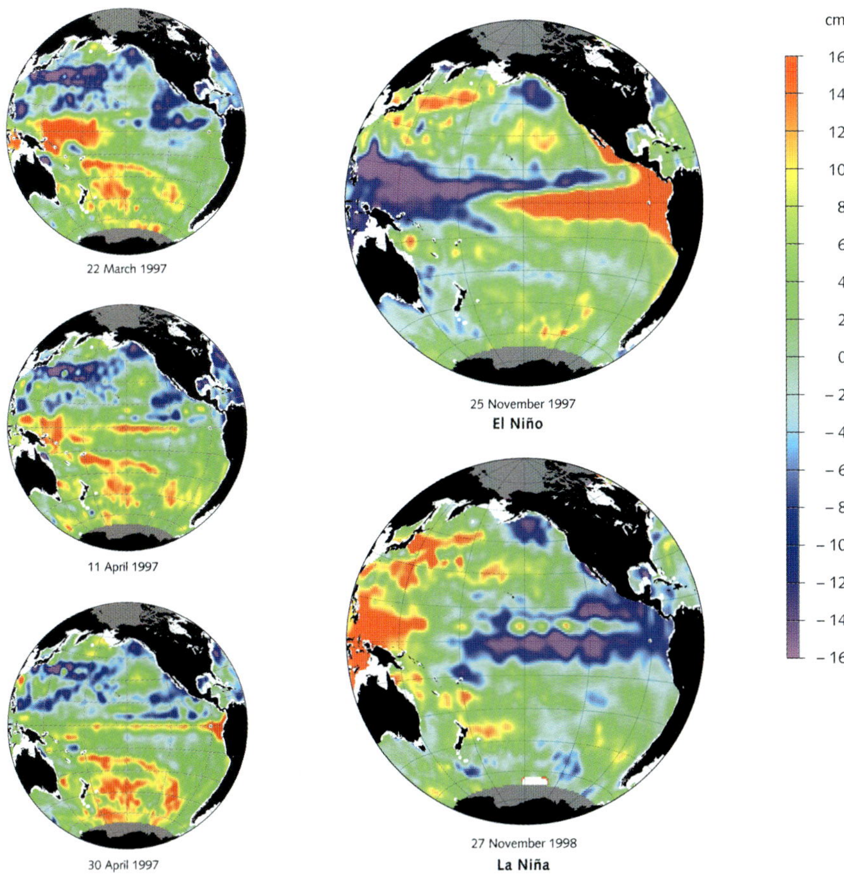

Fig. A

Propagation of a Kelvin wave along the equator observed by the Topex-Poseidon satellite which measures sea-surface height.

The sea-surface height anomalies, in centimetres, are shown here according to the scale to the right of the maps. A positive anomaly of about 16 centimetres (in red) moves from west to east along the equator. The anomaly was near the Indonesian coast on 22 March 1997; it reached the middle of the central equatorial Pacific on 11 April and the coast of the Americas on 30 April.

Fig. B

Sea-height anomalies in the Pacific, observed by the Topex-Poseidon satellite.

The positive anomalies (above-normal sea-surface height) are shown in red; the negative anomalies are in blue-violet, following the scale to the right of the figure.

The contrast between the two situations is marked in the equatorial zone where the positive and negative anomalies are inverted, with a range of sea-height differences of more than 30 centimetres, to the east and to the west.

Images kindly provided by the Laboratoire d'Études en Géophysique et Océanographie Spatiale (Joint Unit of the CNES, the CNRS and the Université Paul Sabatier de Toulouse).

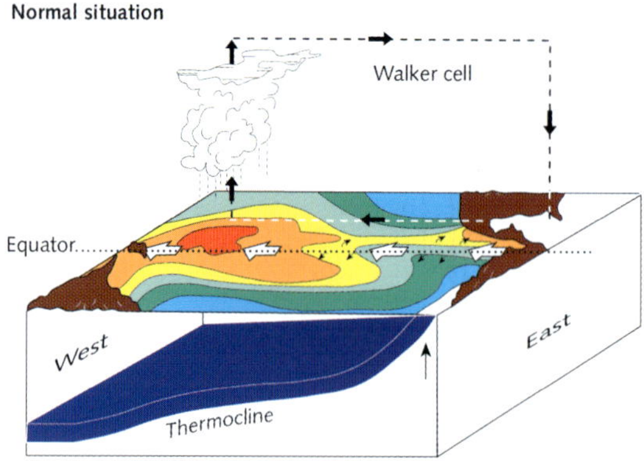

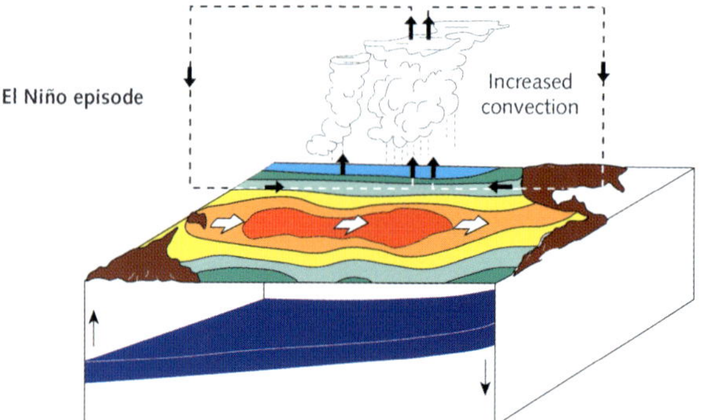

Fig. C

Evolution of the ocean–atmosphere coupling in the equatorial Pacific.
Normally, the trade winds induce upwelling of cool water off Peru and an accumulation of warm water on the western side of the Pacific Ocean, thus raising the sea level between 50 centimetres and 1 metre. Consequently, the thermocline rises towards the sea surface in the upwelling area whereas it is at a depth of some 200 metres near Indonesia. The atmospheric circulation is characterized by strong uplift over Indonesia, which experiences heavy rainfall, whereas the descending air produces arid conditions between Easter Island and the South American continent between Ecuador and northern Chile.
In El Niño years, the slackening of the trade winds causes a displacement of the warm-water mass and the associated atmospheric circulation towards the central Pacific. The thermocline shoals in the west and deepens in the east.

By courtesy of the NOAA/PMEL/TAO Project Office, Dr Michael J. McPhaden, Director.

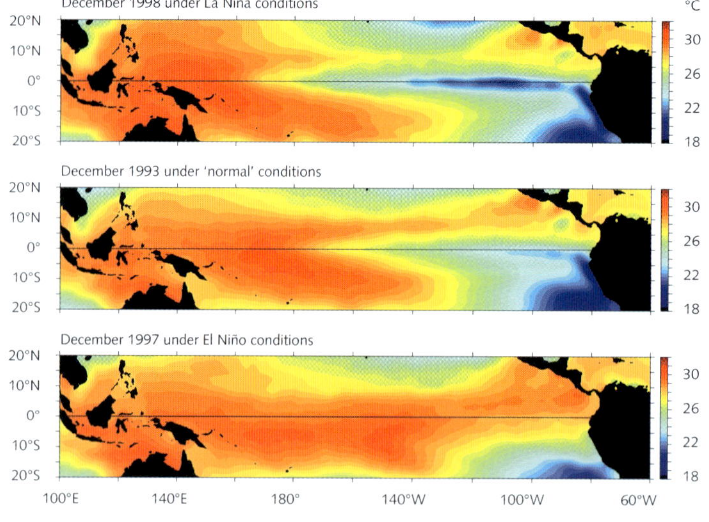

Fig. D

Maps of sea-surface temperature in the equatorial Pacific Ocean.
The temperature colour scale is shown to the right of each map.
Qualitatively, there is a similarity between La Niña and normal conditions, with a thermal minimum along the equator which extends that of the coastal upwelling; nothing similar occurs during an El Niño episode in which a band of warm water is observed along the equator from one side of the Pacific Ocean to the other.

By courtesy of the NOAA/PMEL/TAO Project Office, Dr Michael J. McPhaden, Director.

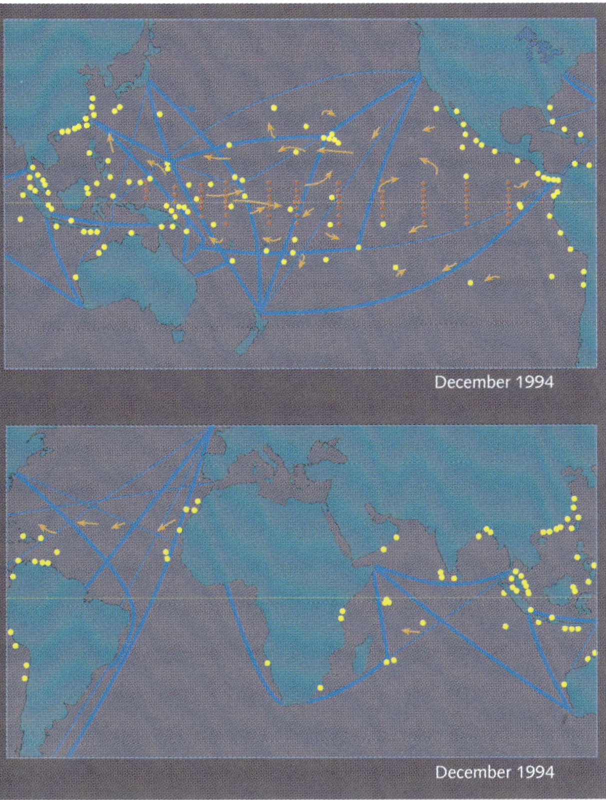

Fig. E
***In situ* tropical-ocean observation system during the TOGA (Tropical Ocean and Global Atmosphere) programme.**

red lozenges = **anchored buoys**
yellow dots = **sea-surface-height-measurement installations on islands and continents**
red arrows = **surface drifting buoys**
blue lines = **tracks of merchant ships carrying out systematic measurements of temperature and salinity**

All the data are transmitted by satellites, which also constitute powerful means of measuring the principal meteorological and oceanographic parameters, including sea-surface height. The observations are particularly dense in the part of the Pacific under consideration, with ENSO as the 'game leader' of the multiannual climatic variation.

By courtesy of the NOAA/PMEL/TAO Project Office, Dr Michael J. McPhaden, Director.

Fig. F
Machu Picchu: *andenes* (terraced crop ponds) built by the Incas to retain rainwater.
Ancient peoples adapted themselves in various ways to the vagaries of climate and the environment. The archaeological vestiges shown in this picture are evidence of the agricultural hydraulic know-how developed by the Incas to deal with such problems and even to take advantage of the heavy rainfall that, much of the time, accompanies El Niño.

Photo: UNESCO/Roque Laurenza.

Fig. G

Hurricane Mitch seen by the GOES satellite on 26 October 1998.

Following the 1997–98 El Niño, which was characterized by very low cyclone activity in the Atlantic in the summer of 1997, the system evolved very rapidly towards a La Niña situation in the summer of 1998, with a spectacular increase in cyclone activity. Hurricane Mitch, which was one of the most violent of the century, is seen here approaching Central America. Its winds exceeded 300 kilometres per hour.

Image kindly provided by NOAA's Satellites Services Division.

Fig. H
Anchovy fishing off the coast of Peru.
During El Niño events, this industrial fishery is greatly depressed by the reduced availability of the resource.

Photos by courtesy of Jürgen Alheit.

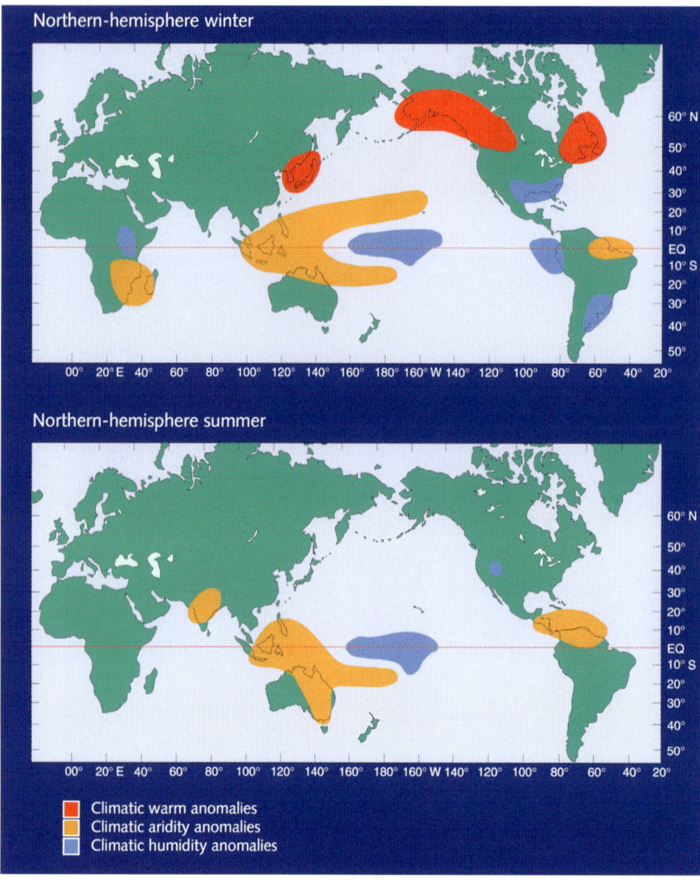

Fig. I

The climatic effects of El Niño on the planetary scale.
Numerous modifications, whether simultaneous or delayed, affect the temperature and the rainfall all over the planet. Among the most constant effects, which are particularly marked during the northern winter, the following may be mentioned:

- drought and heat waves in the western Pacific ('marine continent', Australia), in Central America, in north-eastern Brazil and in southern Africa;
- high temperatures and heavy rainfall in the coastal zone of South America and in south-east Brazil;
- displacement of cyclonic activity from the Indonesian region towards the Hawaii–Polynesia–Cook Islands triangle and reduction of cyclonic activity in the tropical Atlantic;
- weakening of the Indian monsoon.

By courtesy of the NOAA/PMEL/TAO Project Office, Dr Michael J. McPhaden, Director.

Fig. J

***In situ* tropical-ocean observation system during the TOGA (Tropical Ocean and Global Atmosphere) programme.**
One of the anchored buoys in the equatorial Pacific Ocean for continuous measurement of meteorological parameters (the instruments are visible on the buoy) and of oceanographic parameters to a depth of 500 metres (the instruments are attached along a cable suspended from the buoy).

By courtesy of the NOAA/PMEL/TAO Project Office, Dr Michael J. McPhaden, Director.

(a) above: **a house uprooted and displaced.**
(b) below: **two buses bogged down on a road flooded by heavy rains.**

Fig. K
Aftermath of Hurricane Fifi in 1974, in Honduras (La Niña episode). One of the consequences of climate change could be the increased frequency and intensity of such extreme weather. To understand climate processes well, oceanographers and climatologists are unravelling better and better the complexity of the relationship between the ocean and the atmosphere.

Photos: UNESCO/Michel Giniés.

Fig. L
During El Niño periods, large areas of land are often flooded by rain-swollen rivers, forming new lakes or greatly enlarging existing ones.

Photos by courtesy of Miguel Carrillo/El Comercio (Peru).

(a) top: **aerial view, taken 28 February 1998 (El Niño episode), of a transitory lagoon in the Sechura Desert, in the province of Piura, northern Peru.**

(b) bottom: **President Alberto Fujimori during an official visit to the lake the same day.**

4 How ENSO works

OCEAN–ATMOSPHERE COUPLING AND THE BJERKNES HYPOTHESIS

Everything about El Niño illustrates very well the idea of a coupled system such as that formed by the ocean and the atmosphere, each of which has its own particular dynamics. Fluctuation in one of the two components perturbs the other which, in return, accentuates or, on the contrary, stabilizes the fluctuations in the other. Describing the exchanges between the ocean and the atmosphere in the Pacific, Bjerknes indicated:

A change toward a steeper pressure slope at the base of the Walker Circulation is associated with an increase in the equatorial easterly wind and hence also an increase in the upwelling and a sharpening of the contrast of surface temperature between the eastern and western equatorial Pacific. This chain reaction shows that an intensifying Walker Circulation also provides for an increase of the east–west temperature contrast that is the cause of the Walker Circulation in the first place. On the other hand, a case can also be made for a trend of decreasing speed of the Walker Circulation.

We are therefore dealing with a positive feedback system in which it is not known whether it is the ocean or the atmosphere that causes the perturbation. This is the 'two-step' between the Walker cell and its oceanic counterpart, described by Bjerknes, which links the southern oscillation with the east–west temperature gradient in the equatorial Pacific, and is known as ENSO: El Niño – Southern Oscillation, which consecrates the union between the atmosphere and the ocean.

Any index that characterizes one of the two systems also characterizes the other. The southern oscillation index, or SOI, the difference in atmospheric pressure between Tahiti (17°30′S; 149°30′W) and Darwin, Australia (12°25′S; 130°55′E), has the benefit of seniority as a descriptor of the system. Its variation is, generally speaking, a copy, from peak to peak, of that of the sea-surface temperature anomaly in the eastern Pacific: the sea-surface temperature minima correspond to maxima in the southern oscillation index, and vice versa (Fig. 4.1). Also used as an oceanic index is the sea-surface temperature anomaly of the eastern Pacific between 5°N and 5°S and 120°W and 170°W. Now that the Topex-Poseidon satellite allows us to measure the difference in sea level between the two sides of the Pacific to within a few centimetres, this difference could also be used as an index of ENSO.

INDEXES AND ANOMALIES

As is often the case in meteorology, it is the departure from the average, normally over the last thirty years, that is called an anomaly, and which serves as a reference. It is a 'moving' average that varies as a function of climatic change on such a time scale. When the weather forecast is given on television, the presenter rarely omits to indicate whether the temperatures agree with, are higher than, or lower than, the seasonal average, and by how much; in doing so, the presenter is giving us an idea of the anomaly. Usually, the values are above or below normal; this arises because the climate, defined on the basis of a thirty-year average, represents an average of differents types of weather and does not necessarily correspond to the real weather. It is better to speak of an average value than of a norm. Even so, anomaly should not be confused with abnormality. What is aberrant or contrary to accepted laws and theories is abnormal. An anomaly, on the other hand, could be exceptional but 'within the law'! In the course of time, science may transform into simple anomaly, or even normality, what might have been taken hitherto as abnormal. The abnormal does not exist for scientists, who only know the current limits of knowledge and who push these limits outwards as far as possible before coming up against new phenomena at random which the laws of statistics may allow them to normalize. In this sense, El Niño, whatever the anomaly

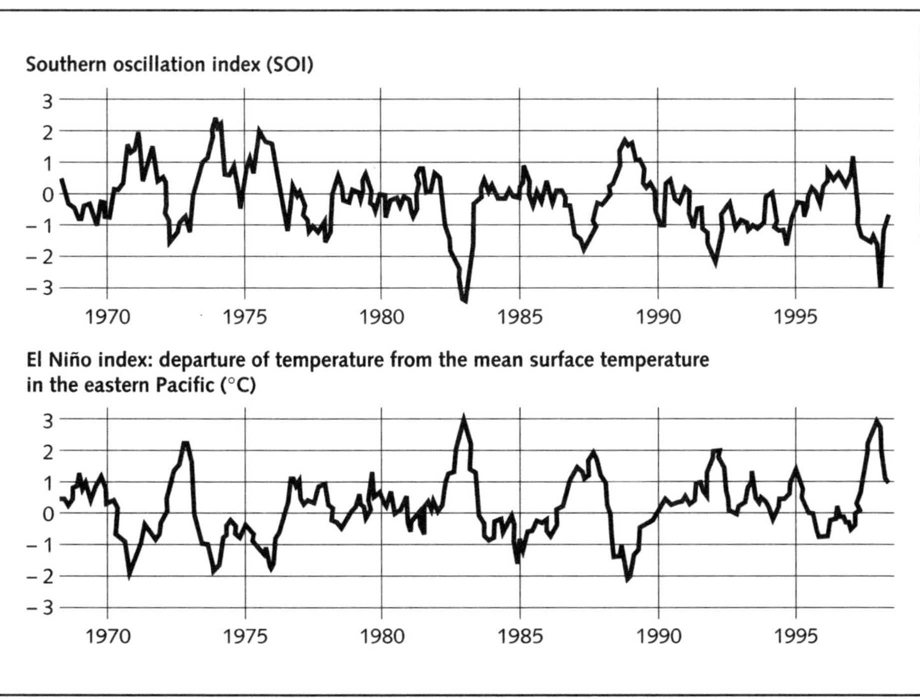

Fig. 4.1
Variation of ENSO as revealed by the southern oscillation index and the El Niño index from 1968 to 1998.
The variation of these two indices shows that the 'ENSO cycle' has a period of two to seven years, with an average of four years. The 1980s and 1990s show increased activity, with five El Niños (1982–83, 1986–87, 1991–93, 1994–95, 1997–98) and three La Niñas (1984–85, 1988–89, 1995–96).
The two most important El Niños of this century (1982–83 and 1997–98) occurred during these fifteen years, as well as a practically continuous El Niño from 1991 to 1995.
The southern oscillation index is the difference in sea-level atmospheric pressure between Tahiti and Darwin.
The El Niño index is the difference in degrees relative to the mean surface temperature in the eastern Pacific.

corresponding to it, is not abnormal; it is a natural component of the climate system. The variation of the southern oscillation index from 1968 to 1998, or more precisely of its anomaly, shows a succession of positive and negative peaks, which well justifies the expression 'oscillation' chosen by Walker to designate this see-sawing between the ocean and the atmosphere. These oscillations are also to be found in the sea-surface temperature which can therefore serve as a climate indicator, with three typical situations called El Niño, La Niña and 'normal', the latter corresponding to a value of the southern oscillation index close to zero (see Fig. D on p. 58).

EL NIÑO – THE WARM PHASE OF ENSO

To describe El Niño, we shall start with the southern oscillation index, without implying that this plays the role of an atmospheric trigger.

When the value of the southern oscillation index decreases, the ensemble comprising the Walker cell and its oceanic counterpart weakens: the intensity of the trade winds, of the equatorial current, of coastal upwelling and of the equatorial divergence decreases (see Fig. C on p. 58). The situation can even become inverse: westerly winds and an eastward-flowing ocean current appear at the equator, and the upwelling disappears. With nothing to retain the warm water accumulated in the Indonesian region, it flows eastwards carrying with it the zone of atmospheric convection, hence the rains (Fig. 4.2). The slope of the sea surface also lessens, rising in the east and descending in the west (see Fig. B on p. 57). This is El Niño, of which the manifestations described in Peru and Ecuador by Pezet or Murphy are simply local signatures. As marxists might have said, it is a 'sudden qualitative leap', which makes one system switch to another.

The contrast with the usual situation is well illustrated by the map of sea-surface temperature (see Fig. D on p. 58): there is no longer a thermal minimum along the equator and, in the eastern Pacific, the temperature increases by 4 to 5 degrees. The exchanges between the two fluids are greatly perturbed. The Walker circulation is dislocated and the flow of the ocean is weakened: the Cromwell Current may disappear, thus halting the equatorial divergence. This situation arose, in the recent past, in 1972, 1977, 1983, 1986, 1992 and 1998, all of which corresponded to positive ocean-temperature anomalies and negative anomalies of the southern oscillation index. The spread of warm water along the equator increases evaporation and the transfer of energy to the atmosphere: the Hadley cell is strengthened and, consequently, the transfer of heat to higher latitudes, especially in the northern hemisphere. The ITCZ, which is associated with the zone of maximum oceanic heat content, follows this movement and is displaced equatorwards.

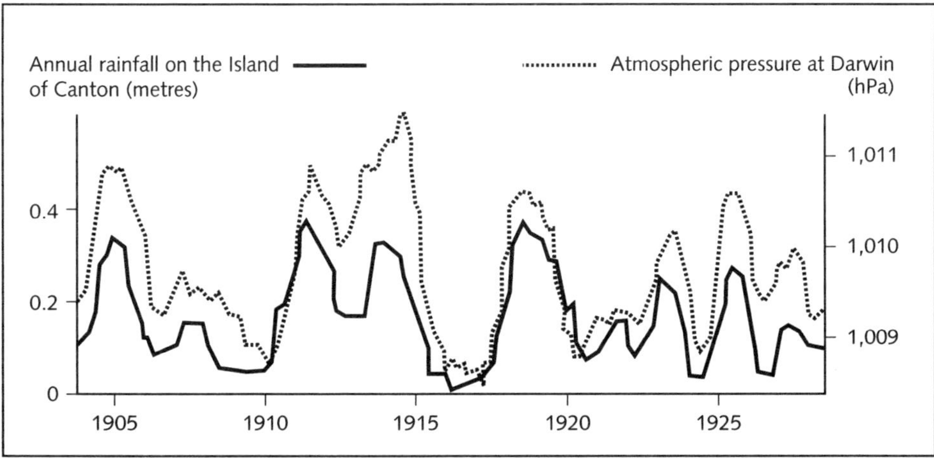

Fig. 4.2
Comparison of the variation in rainfall on the Island of Canton (2°S, 175°E) in the Kiribati archipelago and in atmospheric pressure at Darwin.
In an El Niño year, the 'warm-water pool' and the associated atmospheric convection move eastwards. Rainfall is heavy when the atmospheric pressure is high to the north of Australia, a sign of El Niño.

LA NIÑA – THE COOL PHASE OF ENSO

Even if the maxima of the southern oscillation index are as clear-cut as its minima and correspond to the negative temperature anomalies in the central and eastern Pacific, scientists and the media are much less interested in the cool phase. It is true that, during the last twenty years, they have been much less frequent than the warm phase. The term La Niña, introduced by George Philander in 1986, was popularized only in 1997–98 through phrases such as 'sinister sister of El Niño', 'the infernal couple El Niño–La Niña', 'the *enfants terribles,*' etc. Some people have even proposed other terms, such as 'El Viejo' (the old man, in Spanish), which is not very elegant, or the 'Anti El Niño', not only inelegant but also irreverent, given the origin of the name El Niño.

This lesser interest is, after all, logical, since the climatic status of La Niña is, qualitatively, not different from the so-called normal situation, as the maps of sea-surface temperature show: in both cases, the equatorial upwelling is obvious, with its thermal minimum along the equator. The only difference in this context is that the sea-surface temperature is significantly cooler during La Niña which thus simply accentuates the normal trend, with an intensification in the functioning of the Walker cell: strengthening of the trade winds, accumulation of warm water in the western Pacific, coastal upwelling off the coast of Peru, equatorial divergence. La Niña thus provokes an increase in rainfall over the western Pacific and the 'marine continent' between the Pacific and Indian Oceans. The thermal gradient between the two sides of the Pacific Ocean is strengthened and significantly lower temperatures occur at the equator, hence the description 'cool phase'. La Niña is a phase of maximum activation of ENSO, accentuating the dominant climatic features, whereas El Niño can be considered as a 'breakdown' of ENSO, since the Walker cell in the Pacific and the coupled oceanic gradient collapse. In other words, the two phenomena cannot be set back to back: El Niño destroys or inverts the dominant climatic features, whereas La Niña pushes the system to its limits. The models that explain the ENSO chronology show a succession of warm episodes and cool episodes, with more or less significant lags. Yet it would not be more correct to say that the 1996 La Niña had preceded the 1997–98 El Niño than to say that the 1996 La Niña had followed the 1995 El Niño.

The transition between El Niño and La Niña is often very rapid. This was the case in 1998–99, with a drop in sea-surface temperature in the equatorial Pacific in May–June 1998. A similar thing occurred in the Indian Ocean, causing a decrease in rainfall by half in the southern part of this region and an increase in the northern half. In a few months, Indonesia went from drought and forest fires to heavy rains that caused flooding and mud slides.

IS THERE A 'NORMAL' SCENARIO FOR ENSO?

For a phenomenon to be predictable, it must recur in a similar form sufficiently often for its key factors to be determined. In the laboratory, experiments can be repeated, laws established and their validity tested by varying, in a controlled way, the parameters of the phenomenon under study. Clearly, ENSO cannot be conducted in a laboratory, although the computerized testing of models fed with a growing number of field data is also an experimental approach.

Meteorology advanced thanks to an observation network set up 150 years ago. As this account has reminded us, there is no oceanic equivalent of the World Meteorological Organization to ensure the indispensable continuity of the measurements. Those concerned with the ocean have come from international research programmes with a limited lifetime and from meteorological networks that collect mainly surface temperatures. It is only in the last decade that an effort has been made to set up a thoroughgoing long-term operational ocean-observing network. The Intergovernmental Oceanographic Commission of UNESCO is undertaking to convince its Member States of the merits of continuous ocean observation, not only for climate forecasting on a seasonal scale and on multiannual scales, which are relevant to ENSO, but also for longer-term climate change, particularly that related to the greenhouse effect. It is not a simple matter. First, because of the cost of making such observations, especially since the more fortunate countries consider themselves less vulnerable to climatic change. Then, because political decisions depend more directly on the electoral calendar than on climate change itself. And finally, because politicians, knowing that accurate climate prediction has not yet been established, have no feeling of urgency. It may be noted, however, that it is difficult to achieve the necessary accuracy without a system of systematic observation of the climate as a whole. The meteorological–oceanographic observation network established under the TOGA programme is a start in the establishment of a full network and is becoming the main source of recent progress in our understanding of ENSO.

Comparison of the development of the six warm episodes that occurred between 1949 and 1980 (1951, 1953, 1957, 1965, 1969, 1972) has allowed the elaboration of a so-called 'canonical' (i.e. classical or standard) version of El Niño, with the aim of standardizing the phenomenon so as to predict it by basing its development on a 'normal' year. The play is in five acts covering two years, from June of the year preceding the event until June of the year following it, as follows:

Act 1. *Preliminaries.*The trade winds become stronger at the equator, to the west of the International Date Line, accentuating the slope of the ocean surface and the accumulation of warm water in the western Pacific Ocean. This was, before it was given the name, a La Niña situation. This extreme, and necessarily unstable, situation was considered as being particularly sensitive to atmospheric perturbations.

Act 2. *Take-off.* In October–November of the year preceding the event, gusts of westerly wind occur in the western half of the Ocean,

The forecasting of El Niño – the TOGA programme (see Fig. E on p. 59)

Launched in 1985 for a period of ten years, the TOGA programme (Tropical Ocean and Global Atmosphere) aimed at meeting one of today's major scientific challenges: climate forecasting. TOGA had three objectives:

- to describe the evolution of the coupled ocean–atmosphere system in the tropics and understand the mechanisms causing interannual variations;
- to develop models for the prediction of changes on a time scale of several months to several years;
- to conceive a system of observation and data transmission for the purposes of operational forecasting.

Associated with these international oceanographic undertakings, TOGA was also based on networks:

- of emitter drifting buoys measuring the temperature of the top 20 metres of the ocean and sea-surface currents;
- of seventy moored buoys in the tropical Pacific between 8°N and 8°S; thus, data on the surface meteorological conditions (wind, pressure, air temperature, irradiation) and the oceanic conditions to a depth of 50 metres (temperature and, sometimes, salinity and currents) are gathered continuously and transmitted by satellite;
- of sea-level gauges, to follow changes in sea level;
- ships of opportunity measuring sea-surface salinity and regularly releasing bathythermographs to measure the sea temperature down to a depth of 500 metres.

The TOGA programme has also benefited from the satellite observations of the sea-surface temperature, surface wind speed and direction and ocean topography.

causing sea-surface temperature anomalies in the vicinity of the 180° meridian. In parallel, a positive sea-surface temperature anomaly is observed off the coast of South America.

Act 3. *Peak.* The temperature anomalies increase along the Peruvian and Ecuadorian coasts, culminating between April and June. From thereon, they propagate westwards along the equator; they reach their maximum in the central Pacific between August and December. At the same time, in liaison with the strengthening of the Hadley circulation, the ITCZ moves southwards.

Acts 4 and 5. *Transition and decline.* Starting in the following September, the system reaches its maturity in the first quarter of the following year. The surface temperature returns to its normal value in the west while remaining high in the central and western Pacific. It is at this time that the exchange of heat with the extratropical regions of the northern hemisphere reaches its maximum intensity. The anomaly then decreases and the system returns to normal in the second quarter.

This scenario thus establishes a chronology of events based on the available oceanic and atmospheric observations. It therefore allows a prediction to be made once the precursory signs are detected, but it does not prejudge either the mechanisms or the amplitude of the phenomenon. So the scenario is not explanatory and its predictive capability depends on its stability and its respect for the ocean–atmosphere coupling. This chronology of events, which was published in 1982, was widely accepted as a reliable description of the build-up to a warm episode. But Nature is facetious and sometimes has a cruel humour: just when the event, which by its strength and its consequences, was to become the 'El Niño of the century', no one thought that the 1997–98 episode could steal the show from it. Certain signs might have alerted the scientists, but nobody foresaw it, because it was not classical, and for at least three reasons. First, it was not preceded by a strengthening of the trade winds and the corresponding increase in the slope of the ocean surface; second, it did not develop in phase with the seasonal cycle; and finally, the ocean warming did not propagate from east to west from the coast of South America, but from west to east, starting in the central Pacific!

What then was the uniqueness of the 1982–83 event compared to the long series of events on record? Without going back far in time, it appears that the warm episode of 1940–41 was similar to that of 1982–83. The following episodes (1986–87, 1992–95, 1997–98) were, of course, studied very carefully, thanks to the new observing systems set up in the intertropical zone under the international TOGA programme. Their analysis confirmed the idea that a standard scenario was a mirage; the only constant feature is the local manifestation that gave rise to the name: the warming of the sea off Peru and Ecuador which occurs always in the same season. This finding raises a further question: What is the relation between the seasonal cycle and El Niño? The classical version was the reality between 1950 and 1975. Why did it not remain so thereafter? This in turn raises a final question: What is the relationship between ENSO and long-term climate change? ENSO may be considered, in effect, an autonomous phenomenon with its own dynamics which are independent of the rest of the climate system and on which it imposes its own law. To answer these questions it is necessary to know the processes governing ENSO, which neither the Bjerknes version nor the classical version, which corresponds only to a standardized chronology, allows.

THE DELAYED OSCILLATOR

Bjerknes drew up a coherent framework for the interactions between the ocean and the atmosphere on a large scale. It is the basis of all the proposals for the forecasting of El Niño. But, by itself, it does not indicate the manner in which the phenomenon is triggered, nor how it stops or moves towards the inverse situation, La Niña. Moreover, it is a synchronous system in which all the components of the system vary in phase: the weakening of trade winds off the coasts of South America and across the Pacific, as well as the oceanic warming on the coast and along the equator. It is rather as if, for ENSO, the surface of the ocean were a rigid system like a see-saw made of a plank of wood the ends of which move up and down, or like the beam of a pair of scales. A more realistic system would be the transmission of atmospheric perturbations to the ocean by means of Kelvin and Rossby waves, which move along the equator, are reflected and interfere, the one being able to cancel the effects of the other; this is the model of the delayed oscillator (Fig. 4.3). The great width of the tropical Pacific Ocean, in effect, allows this scenario to run its full course which is determined by the speed of these waves. Thus considered, ENSO practically becomes an oscillation mode specific to the equatorial Pacific and, in this sense, it is predictable. It would certainly be so if we could consider it as being isolated from the rest of the climate system. This is not the case (see Chapters 5 and 6), which explains why there is no 'standard' version of ENSO.

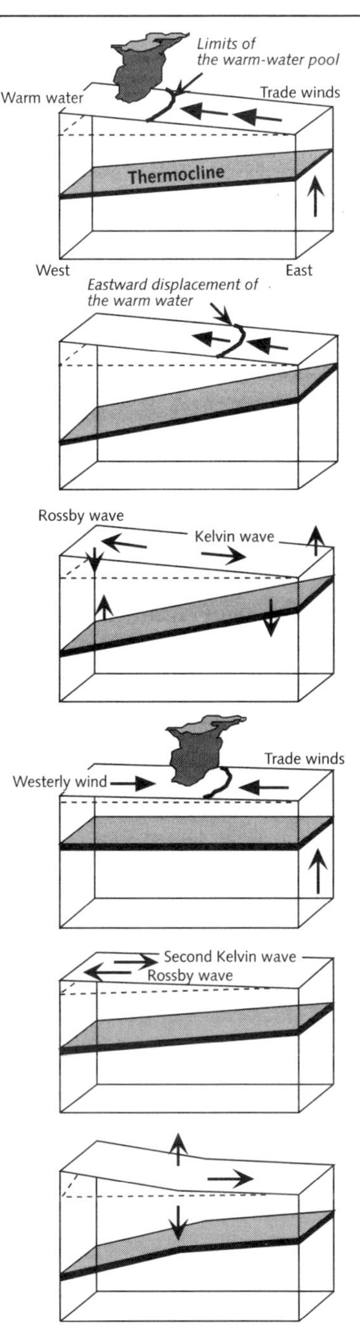

Normally, the trade winds accumulate warm water to the west where the evaporation is intense; this triggers atmospheric convection and rainfall. The thermocline is closer to the sea surface in the east than in the west.

Sometimes, the trade winds weaken in the central Pacific and the warm water moves eastwards.

This perturbation creates an eastward-moving Kelvin wave; it takes two months to reach the coast of South America. At the same time, it induces a Rossby wave which travels westwards at a third of the speed of the Kelvin wave, reaching the western side of the Pacific six months later.

As it passes, the Kelvin wave raises the sea surface and depresses the thermocline, whereas the Rossby wave has the opposite effect. This leads to the El Niño condition in which the sea surface and the thermocline are horizontal. The zone of evaporation is towards the centre of the Pacific.

The Rossby wave is reflected by the ocean border of Indonesia, which generates a second Kelvin wave.

This wave raises the thermocline as it passes eastwards and brings El Niño to an end or, if it is intense enough, initiates a La Niña.

Fig. 4.3
The ENSO cycle explained by the theory of the delayed oscillator.
A symmetric diagram based on an increase in the trade winds would show the birth, life and death of La Niña.

5 On the scale of the planet

A PLANETARY VIEWPOINT

If, at one glance, we could cover the region between the Tropic of Cancer and the Tropic of Capricorn, we should see there three zones of uprising air on the western boundary of the tropical oceans, veritable boilers triggering the ascent of humid air with its accompanying cumulo-nimbus clouds, and three zones of descending dry air underlined by the major deserts. Each ascending or descending branch is thus associated with two adjacent circulation cells, like cog wheels; any change in one of them affects the others. There is thus a link between the three oceans which, however, have their own characteristics. The heating of the atmosphere by the land, which is negligible in the Pacific, is much more important in the Atlantic. Regarding the Indian Ocean, which is subject to the alternating monsoons, it is closed around 25°N, there undergoing the effects of the Himalayan mountain range.

The planetary impact of fluctuations in the atmospheric and oceanic circulations in the Pacific, on a scale of several years, of the ENSO type, implies taking into account variations on other time scales (see Chapter 6). Thus, the atmospheric circulation cells first undergo a seasonal displacement following the insolation maximum, from which comes the alternation between a rainy season and a dry season in the tropics. Variations over a longer period of time have also been found. In the 1920s, the mean pressure of the Azores anticyclone increased by 7 hectopascals, whereas it decreased in the South Pacific. The latter modification accompanied a displacement of the high-pressure centres from 115°W in 1915 to 105°W in 1945, and which is now stabilized at around 95°W. This has led to a decrease in the frequency of hurricanes in the Caribbean, as well as a strengthening of the trade winds off the coasts of Venezuela and Colombia. This reduced frequency of cyclones brings about a decline in the prevailing south and westerly winds that bring the humidity of the Pacific Ocean to Central America: this has led to a 70% decrease in rainfall in Costa Rica.

With hindsight it is possible to analyse the planetary climate changes that accompany the movement of the 'see-saw' of the southern oscillation and that characterize these warm episodes (El Niño) or cool episodes (La Niña). Those changes that are linked to El Niño are easier to demonstrate because this event is often revealed by a marked change in the climate, whereas during La Niña there is only a strengthening of familiar features.

IN THE TROPICAL ATLANTIC

The tropical areas of the Pacific and the Atlantic have some features in common:

- → the presence of an intertropical convergence zone towards which the trade winds of both hemispheres blow;
- → atmospheric circulation is via the Hadley and the Walker cells;
- → same ocean-temperature distribution, with upwelling along the coast and at the equator;
- → similar oceanic circulation.

Their totally different basin morphology differentiates them in respect of their relationship with the atmosphere and of their climate variability. The Pacific Ocean is huge: 17,000 kilometres wide at the equator; it is cut off from the Arctic by the Bering Strait and the Aleutian Islands. The Atlantic Ocean is much slimmer: only 6,500 kilometres wide at the equator; it is completely open to the Arctic, as well as to the Antarctic. This north–south extent confers on it a major role in long-term climate variations (from a decade to a millennium), but we do not deal here with this long-term aspect.

In spite of these similarities, the Atlantic does not have the exact equivalent of the ENSO system of the Pacific, for two reasons: the narrowness of its basin and the impact, on the eastern side, of the African continent which provokes a monsoon in the Gulf of Guinea. Between the zone of convection over the Amazonian forest and that over the forest of equatorial Africa, the Walker cell has hardly enough space to be able to operate. The ascending limb is over the Amazon region and the descending limb covers the whole of the Atlantic from north-east Brazil to south-western Africa, and the climatic variations on either side are in phase. This is the opposite of the situation in the Pacific Ocean in which the two sides are in opposite phases. Since the distance from one side of the Atlantic to the other is relatively short, a mechanism analogous to that of the 'delayed oscillator' cannot become established to generate the oscillations analogous to those of ENSO in the Pacific. In effect, the surface-temperature variability is two times lower in the equatorial Atlantic than in the Pacific.

The interannual variations in the warming of the Atlantic do not therefore come principally from an eastward displacement of warm water along the equator, as in the Pacific, but from the Atlantic Ocean's response to an atmospheric forcing: the changing position of the ITCZ. The variations in the sea-surface temperature linked to the displacement of the meteorological equator are at least as important as those along the (geographical) equator. There is a good correlation between the temperature anomalies to the north and to the south of the equator (signatures of anomalies in the position of the ITCZ) and rainfall variations over north-east Brazil. It is the southward displacement of the ITCZ during the northern-hemisphere winter that generally brings rain to north-east Brazil, a region known for its aridity, with extreme droughts called '*secas*' locally. But does this mean that the hydrological regime in north-east Brazil is independent of El Niño? Certainly not; remember that 1877, an El Niño year, witnessed a total rainfall deficiency. In effect, if there are no real Atlantic El Niños, marked El Niño/La Niña episodes in the Pacific trigger similar events in the Atlantic, but in the opposite phase. As an example, let us detail the sequence of events in the Atlantic in 1983–84 linked to the powerful El Niño of 1982–83 followed by a cool episode in the Pacific in 1984. In contrast to the Pacific, 1983 was a particularly cool year in the Atlantic; equatorial upwelling there was very active, whereas, in 1984, which was a warm year in the Atlantic, it practically disappeared. During an El Niño (1983), the Pacific convective zone is displaced eastwards, with the easterly and the westerly winds converging towards it. To the east of this zone, the easterly winds are strengthened and, by continuity, the Atlantic trade winds are attracted by the uncustomary proximity of this zone of convergence. This strengthening activates the small Atlantic Walker cell and its descending limb over the Atlantic Ocean and north-east Brazil which consequently experienced a dry period. The opposite situation arose in 1984: La Niña dominated in the Pacific and, at its eastern end, the descending limb of the Walker cell became more an area of divergence of the winds which had the effect of repelling the Atlantic trade winds. These weakened; the Atlantic passed over to the warm phase favouring the southward descent of the ITCZ, which in turn favoured rainfall in north-east Brazil but also in the desert areas of the African coast, owing to the narrowness of the Atlantic basin.

To summarize: ENSO influences the Atlantic, especially north-east Brazil. In this region, most of the El Niño episodes coincide with dry years; however, there are also dry years, and even very dry years, that are independent of these El Niño episodes. A statistical approach shows that the Pacific El Niño accounts for only 10% of the rainfall variation in north-east Brazil. It is not much, but enough to cause a shift from ordinary dryness to extreme aridity. The correlation between the ENSO index and the rainfall over north-east Brazil is much weaker than that between this rainfall and the position of the ITCZ.

For convenience, we speak of Atlantic El Niño and La Niña, but these episodes do not correspond to a specific Atlantic oscillation; they are governed by ENSO. To a marked El Niño situation in the Pacific there corresponds a situation of the La Niña type in the Atlantic, and a warm episode in the Atlantic occurs when the Pacific enters a cool phase. This relation is not at all systematic, since only three episodes of the El Niño type were recorded in the Atlantic in the last quarter of a century: 1963, 1968, 1984.

This phase contrast between the two oceans also shows up in a spectacular way in the frequency of cyclones. In the warm period in the Pacific, cyclone activity is reduced or even zero in the Atlantic. This is due to the modification of the path of the high-altitude air currents, the jet stream (which blows eastwards at an altitude of about 10 kilometres and adds an hour to the flight time of jet aircraft crossing the Atlantic from east to west). A shear is produced between the jet stream and the trade winds blowing westwards, which inhibits the vertical extension of the convection that is a necessary condition for cyclone growth. By contrast, in a La Niña period, cyclonic activity increases over the Atlantic: on 24 September 1998, exceptionally, three cyclones could be seen simultaneously.

THE RELATIONSHIP BETWEEN EL NIÑO AND THE INDIAN OCEAN MONSOON

In the Indian Ocean, the dominant climatic signal is seasonal: it is the alternating monsoon regime driven by the seasonal pressure variations over the Asian continent. The word monsoon comes from the Arabic word *mausim*, which means 'season' or the 'wind of spices and fecundity'. Without taking advantage of the monsoons, the Romans would never have been able to maintain, from the beginning of the Christian era, close commercial relations with India and China. The spices (pepper, nutmeg, cloves, ginger, cinnamon) coming from the islands of Asia were familiar products to the Romans. From China, from Ceylon, from India, the merchandise was carried by Indian or Persian ships, taking advantage of the monsoon, right up to the entrance to the Red Sea. From there it was carried overland to Antioch or Alexandria, then across the Mediterranean to the port of Ostia. A whole year was needed for this voyage, but this was a lot less than the time the Portuguese took via the Cape of Good Hope, which tripled the distance. The relative slowness of the journey was compensated by the size of the fleet which exceeded a hundred ships and even allowed the Romans to establish a trading post at Virampatnam near Pondicherry in south-west India. This maritime route overshadowed the land route taken by the caravans controlled by the Parthians. After the fall of the Roman Empire and the spread of Islam, the Indies route fell, in the 7th century, into the hands of the Arabs who established trading posts in the Indies and in China. In the 15th century, the closing of the overland Silk Route by the Ming dynasty gave the Moors the monopoly of the trade between the Far East and Europe. Without the monsoon, these navigators would not have been able to cross the Sea of Oman aboard high-speed sailing vessels which allowed them, in the summer, to sail before the wind to India to seek the spices of the Malabar coast, and return in winter. It was not until 1786 that, before all others, Antoine d'Entrecasteaux succeeded in sailing from Ceylon to China against the monsoon.

In 1686, the English astronomer Edmond Halley showed, with good reason, that this inversion of wind direction was due to the temperature difference between the ocean, where the temperature range is small, and the continent, where it is very large (Fig. 5.1). More precisely, it is the pressure differences caused by the temperature variations that are the true cause. Whereas, over the ocean, the atmospheric pressure varies little, the differences between summer and winter over the continent are considerable. During the northern-hemisphere summer, the Asian land mass heats up considerably, creating a centre of very low relative pressure towards which the oceanic air, of high humidity and a source of welcome rainfall, is attracted. The arrival of this south-west monsoon is celebrated because it marks the end of a period of strong heat and drought. With the Himalayas forming a barrier, the northern parts of India receive abundant rainfall, the Cherrapunji region receiving more than 11 metres per year. These rains persist until the arrival of the northern-hemisphere winter. The winter cooling creates over the centre of Asia a vast and extremely powerful anticyclone which generates, from the continent towards the ocean, a kind of gigantic and very dry 'land breeze'. Over the ocean, the air becomes charged with humidity, to the benefit of northern Australia.

In this account, we started from the dreadful 'failure' of the 1877 monsoon, and from the concern of Blanford and of Walker to establish relationships that would allow its prediction. The southern oscillation then led us to the Pacific, the main driving force behind ENSO, thus relegating variations in the Indian monsoon to the edge of a phenomenon centred on the Pacific, then considered as the real conductor of the interannual seasonal variability. A statistically significant correlation exists between the southern oscillation index and the rainfall anomalies over India during the summer monsoon; it associates El Niño (weak index) with deficient monsoons and La Niña (strong index) with strong monsoons. The result is a variation of more or less 20% in the rice harvest. Curiously, the abundance of the monsoon rains is more strongly correlated with the sea-surface temperature of the central equatorial Pacific Ocean than with the southern oscillation index.

There is then a link between the two phenomena. It was in this sense that the international TOGA programme (see box, p. 72) was organized under the auspices of the World Meteorological Organization, the Intergovernmental Oceanographic Commission and the International Council for Science, and which laid a basis for interannual climate

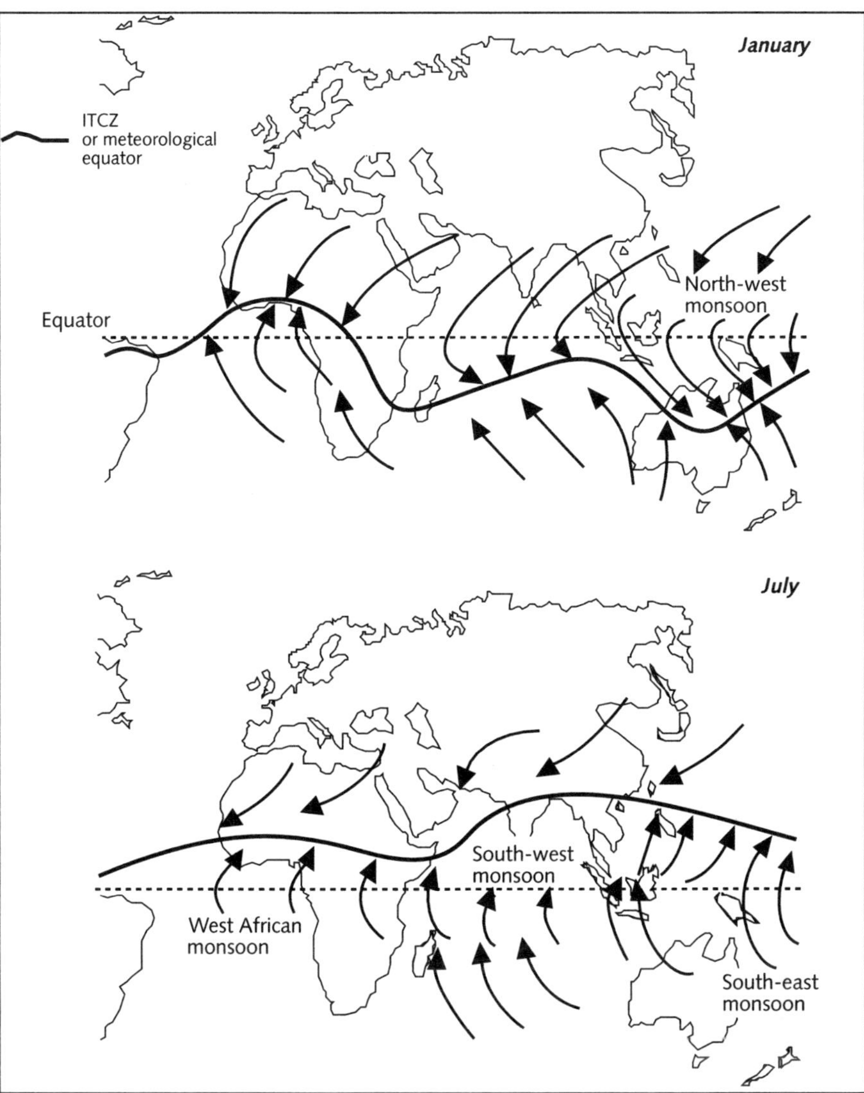

Fig. 5.1

The monsoon phenomenon.

The monsoon is characterized by the seasonal reversal of the winds, especially in the Indian Ocean. The trade winds, drawn by the migration of the intertropical convergence zone, change direction when crossing the equator at which the Coriolis force is inverted.

During the northern-hemisphere summer, the meteorological equator and the associated winds migrate to the north of the geographical equator, following the Sun's seasonal movement. These winds pick up humidity and heat over the Indian Ocean, thus ensuring high rainfall, especially in India and Indonesia. The monsoon phenomenon recalls the daily alternation: land breeze–sea breeze, since it accentuates the thermal difference between the continent and the ocean; this is especially marked in the Indian Ocean with the Himalayas to the north.

prediction. This programme was largely dedicated to the tropical Pacific Ocean. This was a logical step, given the necessity of having available, over a period of time sufficiently long relative to the variations in ENSO, the measurements needed to understand the mechanisms. From this Pacific-centred point of view, the results have been worthwhile, but how well does this stand up to the declared objective: multiannual climate forecasting on a planetary scale and not just on the two sides of the Pacific? Or, to put the question in a different way: what does an improved knowledge of the functioning of ENSO in the Pacific bring to the prediction of the monsoon? The statistically significant correlations between ENSO and the variability of the monsoon have not been improved as a result of TOGA, nor are they strong enough to serve as a basis for a prediction tool. In the period 1870–1991, 22 years with a deficient monsoon were recorded, of which only 11 corresponded to an El Niño episode. In contrast, of the 18 years in which the monsoon was strong, only 7 were La Niña years. This relative setback recalls that of the complicated prediction formulas of Walker, in spite of our much better knowledge today of the physics of these phenomena and the progress of modelling. The evidence cannot be denied: ENSO is far from explaining all the sources of variance of the monsoon on an interannual scale. It is not enough to forecast ENSO so as to be able to forecast climate variability in general and the monsoon in particular.

Regarding the interannual climate variability for the Atlantic and the Pacific, we are almost exclusively interested in the ocean and the atmosphere. It is impossible to reason in that way for the Indian Ocean, which is only 'half an ocean' closed at about 25°N by the imposing Asian land mass which thus plays an important role in the climate game. It imposes on the system its own rules and its characteristic scales of variability. The game goes from two players to three, which makes it different and more complex. To be able to understand it properly, it is necessary to know how the atmosphere, the ocean and the continents communicate amongst themselves. They do so by exchanging heat, humidity and momentum. The principal points of communication are the interfaces between the ocean and the atmosphere, on the one hand, and the continents and the atmosphere, on the other. The atmosphere serves as the liaison, the messenger, between the tropical ocean and the continent: its movements and its properties (the message it carries, hence the monsoon) are tied to these two poles (its 'informers') and to their variability. It is a good messenger because its response time is short: it reacts rapidly to the changes in its 'informers'. As for the Atlantic, the interannual variations of the sea-surface temperature of the tropical Indian Ocean are much weaker than those of the equatorial Pacific Ocean. But, unlike the Atlantic, they move in the same direction throughout the area under consideration, following closely the changes in the ENSO index. On a multiannual time scale, the Indian Ocean is only an annex of the Pacific Ocean following the rhythm of the changes in the reservoir of warm water in the western Pacific. The variations in the intensity of the monsoon rainfall are moreover much better correlated with variations in sea-surface temperature of the equatorial Pacific than with those of the Indian Ocean. These correlations are, however, insufficient for effective prediction, because of the third player: the Asian continent, which is subject to extratropical climatic influences which force us to go outside the once-prevailing regional view of the monsoon and to take into account, for prediction purposes, the climate system as a whole with all its various scales of variability. The monsoon and ENSO have in common the Indo-Pacific pool of warm water that supplies, by convection, the Walker cell in the Pacific and the fluxes from the monsoon. This is the common feature that explains the significant correlations between the ENSO indexes and the monsoon. Such a correlation underlines the existence of a relationship but does not necessarily express a cause-and-effect relationship. We come back to the question of the chicken and the egg, and it must be admitted that, if the fluctuations in ENSO are reflected in those of the monsoon, it can also be said that the variations in the latter have an influence on ENSO. Compartmentalizing the climate system, with ENSO on one side and the monsoon on the other, is useful for analysing the physics of these phenomena. Nevertheless, this reductionist approach, which is a necessary stage in the scientific process, is not sufficient to resolve the variability of the climatic system, which also, and perhaps even more, depends on the interaction between these two systems, as well as on the influence of the extratropical regions.

In 1953, Normand wrote:

> It is quite remarkable that Indian monsoon rains be more closely linked to events following it than to events preceding it. It is quite in keeping with this that the Indian monsoon rainfall has its connections with later rather than earlier events … Unfortunately for India, the southern oscillation in June–August, at the height

of the monsoon, has many significant correlations with later events and relatively few with earlier events ... The Indian monsoon therefore stands out as an active, not a passive feature in world weather, more efficient as a forecasting tool than as an event to be forecast. On the whole, the Walker worldwide survey ended offering promise for the prediction of events in other regions rather than in India ...

In other words, a return to square one, or almost so after the detour via ENSO and the TOGA programme!

PLANETARY TELECONNECTIONS: THE NORTH PACIFIC AND THE REST OF THE WORLD

The 'teleconnection' reflects the links that exist between climatic anomalies a long distance from each other. ENSO is itself a teleconnection linking the atmospheric pressure anomalies between Tahiti and Darwin via the Walker cell. Step by step, as we have seen previously, one can also speak of teleconnection between ENSO, on the one hand, and the Indian monsoon or the rainfall over north-east Brazil, on the other hand. Teleconnections necessarily correspond to statistically significant correlations between the anomalies concerned. Yet it is not enough that there be such correlations between two phenomena for us to speak of teleconnection: an explanatory mechanism is needed.

The term 'teleconnection', introduced by Anders Ångström in 1935, was soon put aside somewhat because of misuse: one can always look for correlations between anything and anything, and it is tempting to deduce automatically from them that they correspond to a physical connection, even if unknown, whereas the correlation could be purely fortuitous. A simple correlation could take the place of reasoning, so anyone using this word could become suspected of intellectual laziness. We have, in the preceding paragraphs, but without using the word, analysed the teleconnections between ENSO and the tropical areas of the Indian and Atlantic Oceans: those that are mainly relayed by the Walker cells.

This name has been reserved for the interactions with the extratropical regions. The transmission belt for these perturbations is then the Hadley cell, which operates in the meridional plane (north–south). An El Niño event that provokes a spreading of the warm-water pool across the whole width of the equatorial Pacific also provokes the descent of the ITCZ towards the equator, an increase in atmospheric convection, hence an activation of the Hadley cell which progressively transfers more energy towards the high latitudes. The air temperature in the troposphere increases, as therefore does the thermal gradient from the equator to the high latitudes.

IN THE NORTH PACIFIC

The transfer of energy via the Hadley cell towards high latitudes also occurs in a wave form, which periodically increases or decreases the atmospheric pressure northwards. During El Niño this has the effect of, for example, reinforcing the area of low pressure in the North Pacific (the Aleutians) and of facilitating the entry of marine air to the north-western United States and Canada, which then experience mild and humid winters. The supply of supplementary energy in the Hadley cell is also evacuated partly in the upper troposphere by the strengthening and eastward extension of the subtropical jet stream which becomes established in the zone of temperature gradient (also strengthened), limiting the Hadley cell on its northern side. This jet stream is accompanied by gusts of wind and storms in winter over California and Mexico. It is also its eastwards extension that limits cyclone formation over the Atlantic. During La Niña, the subtropical jet stream weakens, giving rise to drought in Mexico and the Gulf of Mexico, as well as to an increase in the frequency of Atlantic cyclones.

The example of the southern United States shows, however, that many precautions are required before linking a modification in the temperate zone to the southern oscillation. The severe winter experienced in the south-eastern United States in 1976–77 was thus attributed to El Niño. In Florida, oranges covered with ice were seen hanging from the branches of orange trees. During the 1982–83 episode, on the other hand, the winter was the mildest in the last twenty-five years: energy consumption decreased and the cereal harvest was superabundant! This shows the necessity for establishing teleconnections on a physical basis.

The transfer of energy towards high latitudes also occurs via the ocean, which is much slower and, because of that, less well known. Thus, the warming of the Pacific around 40°N was detected in 1993. Many oceanographers agreed that this was one of the effects of the 1982–83 El Niño! In the section 'El Niño – the warm phase of ENSO' (see Chapter 4), we only recounted the story of the intertropical zone. But the story does not end there. Thus, when the warming reaches the coast of South

America, as in 1982–83, the warm Kelvin wave travels along the coast of Central America then North America, to reach, three months later, 40°N latitude. The elevation of the sea level triggers a Rossby wave which travels westwards at depth, very slowly, since its speed is inversely proportional to the square of the latitude. In 1991–92, this wave reached the area of the Kuroshio and was pushed northwards, leading to a warming in excess of 1 degree which lasted until 1992–93. This perturbation has an effect on the atmosphere, hence on the meteorological conditions. A new stone in the edifice of climatic complexity: might there be a relationship between the flooding of the Mississippi in 1993 and El Niño in 1982–83?

THE REST OF THE WORLD

Beyond the foregoing, it is difficult to use the term 'teleconnection' in the physical sense we have given it. Present knowledge does not allow us, in effect, to link climatic events in Europe and the Middle East to the southern oscillation. The floods in western Europe (25 dead, $200 million worth of damage) or the cold spell in the Middle East (65 dead, $50 million worth of damage) should not be counted as El Niño effects, as was done for the 1997–98 episode, and for which no link with El Niño has so far been proved. There could, of course, be some more or less significant and fortuitous correlations with the rest of the world, which it would be wise not to interpret as consequences of ENSO: during an El Niño period, any climatic anomaly anywhere in the world, whether beneficial or catastrophic, should not be systematically laid at the door of El Niño which, in spite of its name, should not be made a scapegoat.

6 Can ENSO be predicted?

SUCCESSES AND FAILURES OF FORECASTING

Scientists have long been tempted to forecast El Niño, but, inevitably, with a probability of error inasmuch as the announcement of an El Niño or a La Niña must be made several months in advance to be useful in socio-economic terms. As an Australian brochure published in 1991 pointed out: 'We can express the chances of getting more than a certain amount of rain in any month. But a 66% probability of getting more than, say, 100 mm of rain still means that one year in three we could make a wrong decision.'

The history of attempts to forecast El Niño illustrates the difficulty of the exercise, given the caprices of a chaotic climatic phenomenon. It is all recent history, since it only started in 1982 with the classical version which was immediately contradicted by the 1982–83 event which 'failed'. One example among others: a teacher at an American university predicted a good 1983 wheat harvest, based on a strong correlation between El Niño and the wheat harvests in Illinois, and the press made much of it; unfortunately, the harvest was only half its usual level, which did not enhance the reputation of researchers.

Then, with Stephen Zebiak and Mark Cane of Columbia University, the scientists met with success, announcing the 1986–87 El Niño several months in advance, using a simple ocean–atmosphere model. Certain politicians followed in their footsteps: the Ethiopian Government, convinced by the meteorologists, adapted agricultural production to an El Niño year which is characterized by a strengthening of the short rainy season (mid-February to mid-May) and low rainfall during the long rainy season (June to September). The government encouraged the farmers to sow and fertilize to a maximum for the short rainy season as a way of compensating for the losses that would result from the summer dryness. For the latter, it advised limiting the area to be seeded and to use seed of fast-growing plants. This same model also predicted the start of the 1991 event. The governor of the State of Cearà in north-east Brazil believed the prediction and took steps to limit the effects of the drought. The farmers were advised to use seed of fast-growing plants in a dry environment, and preventive water-saving measures were taken in the capital, Fortaleza. Although rainfall was down by a third, the cereal production reached more than 80% of that of a year with normal rainfall. Comforted by these successes, the authors suggested in 1991, not without reason, that the mechanism of El Niño was simple:

> The degree of forecasting skill obtained despite the crudeness of the model is telling. It suggests that the mechanism responsible for the generation of El Niño events and, by extension, the entire ENSO cycle, is large-scale, robust and simple. If it were complex, delicate or dependent on small-scale details, this model would not succeed.

Nature perhaps felt offended by this simplism, for it took its revenge at the first opportunity. The disappearance of El Niño, expected at the end of 1992 by the scientists, disappeared! The warm event lasted almost four years, a scenario that is anything but classical and hardly compatible with the hypothesis of the delayed oscillator. Even if offshore Peru and Ecuador experienced three light El Niños, at the beginning of 1992, at the beginning of 1993 and at the end of 1994, the open-ocean temperature between 150° and 160°W remained, in contrast, constantly above average and we can speak of a continuous El Niño from 1990 to 1995. The US National Weather Service, whose long-term forecasts stress the temperature changes in the Pacific, recognized its difficult mission: 'The El Niño now developing came as something of a surprise … It is the third in four years and El Niño forecasts didn't see it coming until late summer 1994.' This model was also found wanting in 1997 and the hope it had inspired prevented sufficient attention being given to signs that were detectable six months in advance and to the results from other models that took better into account the complexity of the relations between the ocean and the atmosphere. No, Nature does not like simplification and, in contrast to what has just been said, it is complex, refined and sensitive to small details. Nevertheless, the 1997–98 event was a godsend to the scientists because it was the first of such intensity to have had the benefit of so complete and dense a network of observations, allowing its development to be followed day by day. Having failed to foresee its triggering phase, however, what lessons can be drawn from it?

CAUTIOUS OPTIMISM

The results from models coupling general atmospheric circulation and ocean dynamics show that the eastern tropical Pacific is 'predictable' one year in advance if the models are adjusted periodically using the measurements provided by the observation network. The comparison between the change in sea-surface temperature of the equatorial Pacific and the change predicted by the models shows, retroactively, that the warming that started in early 1997 and ended at the end of the same year had been foreseen as early as November 1996. It was their scrupulous respect for the experimental method that inhibited the scientists; they waited until the results of the simple model, which had worked well hitherto, had been completely disproved by the observations. As from April 1997, the surface-temperature forecasts for the following months proved to be satisfactory. Was this a definite advance or might it only have been a new turn of a phenomenon that shows, each time it occurs, that it is unforeseeable? To an embarrassing question, a prudent answer: the models have undoubtedly progressed; they can take into account all the complexity of the ocean–atmosphere system, and the means now in place have no common denominator with those that existed prior to TOGA.

These models have thus provided a satisfactory scenario of what was to happen in 1997, even if they underestimated the amplitude. At the same time, the observations from the network showed that, in the western Pacific, the transfer of heat from west to east along the equator had already started in September 1996 at a depth of 150 metres without any apparent trace at the surface. In other words, the scenario had already been under way for some months when the modellers delivered their first forecasts in November 1996. The sea-surface temperature anomalies, the signature of El Niño, appeared only in March 1997. Would it have been possible, with these same models, to have foreseen them a year earlier? For the moment, the answer is negative and it is not impossible that it will remain so, to the extent that a slight modification of the initial conditions can change the evolution of the system. Now, no model can do any more than extrapolate, with sophisticated numerical techniques, the dynamics of the system. If the system hardly changes, we speak of a 'blockage', and the forecaster can do nothing.

How then to foresee how a situation described as normal and apparently stable in the equatorial Pacific will evolve into an El Niño event before we have detected the forerunners? Or, even more tricky, how to foresee during the run-up to an El Niño that the system will evolve some months later into a La Niña, or even the inverse? This takes us back to the unsolved problem of El Niño triggering mechanisms and the various indicators thereof. Every event has its own character that makes it incomparable with any other. Should we deduce that there are different triggering mechanisms? Maybe, but at least we should admit that their success depends largely on the climatic conditions at the time of their appearance: seasonal variations, decadal variations, atmospheric background noise, the influence of the extratropical regions which link El Niño inextricably to the climate system as a whole, hence the difficulty of forecasting it.

Thus, the delayed oscillator concept, even if it describes the way in which the ocean propagates, in the form of waves, the mechanical energy received from the atmosphere, is no doubt in turn disqualified, owing to an excess of simplism. It is a sort of compulsory ice-skating figure imposed on Nature, which, like ice skaters, prefers to express itself in the optional figures. The classical version of 1982 had a unique chronology that followed the calendar. That of the oscillator corresponds – and that is an advance – to a unique process indifferent to the calendar, but which reflects its own mode of oscillation of the ocean basin with a period of about eighteen months. Three successive waves from 1991 to 1995 clearly do not fit into this scheme. The observations from 1997–98 validated the underlying process, the propagation of Kelvin waves, but it did not conform to the ideal sequence of the delayed oscillator. In effect, on the western side of the basin, the atmosphere was the site of oscillations with a period of forty to fifty days, inducing westerly wind bursts. Each burst was accompanied by a 'downwelling' Kelvin wave the progress of which was followed by the Topex-Poseidon satellite and by the observation network set up under the TOGA programme. Such events occurred in December 1996, February, May, August, October and November 1997. This succession of Kelvin waves did not allow the delayed oscillator scenario to be brought into play, since, every two or three months, the 'game' was restarted by a new westerly wind burst. When, starting in May 1998, the situation was turned around into a mild La Niña, no one could say whether this was according to the delayed oscillator scenario following a westerly wind burst.

At the risk of being contradicted by the next event, El Niño, or at least the appearance of the sea-surface temperature anomalies that characterize it, and their evolution, can be foreseen several months in advance thanks to coupled ocean–atmosphere numerical models and to *in situ* observations which allow regular recalibration of the models to the real ocean. But this does not respond to the expectation of a forecast of the amplitude of the climatic perturbations in one part of the world or another. The statistical relations between the ENSO indicators and the weather in the tropics are not strong enough to constitute operational forecasting tools. The examples of north-east Brazil and the Indian monsoon have taught us a lesson in this regard. Thus, with reference to the southern oscillation index and the sea-surface temperature, the events of 1982–83 and 1997–98 are of comparable importance; they have not themselves had the same consequences, however. In 1997–98, Australia certainly had a rainfall deficit, but it was not at all comparable to the severe drought of 1982–83, and the monsoon over India was practically normal in the summer of 1997. In contrast, Kenya and southern Somalia had, from October 1997 to January 1998, the most abundant rainfall since at least 1961. The prospect of forecasting regional climatic variations linked to ENSO, based on simple indicators such as atmospheric pressure or ocean-surface temperature, corresponds to an idealization of the phenomenon 'dictating' climate variability without being perturbed by the rest of the climate system. This idea has to be abandoned. El Niño and La Niña must conform – to the inevitable complexity of the climate system which cannot be reduced to a few simple recipes and described by a few indicators, however good they might be as integrators. To foresee climate impacts, there is no other solution than to have recourse to coupled ocean–atmosphere models, which have allowed us, retroactively, to foresee the 1997–98 event a few months in advance and which are being tested in the forecasting of climate variability anywhere in the world. These models make ENSO commonplace, its only unique feature being that the amplitude of its variability is particularly marked, hence perhaps more easily forecast in the tropics than elsewhere. It thus becomes inappropriate to speak of the 'consequences' of El Niño since it is itself the result of this variability which it no more determines than does the monsoon. All these events are interactive, and the southern oscillation index is an indicator of the variability of the climate system as a whole. Faced with this complexity, to speak of El Niño triggering mechanisms is a convenience allowing us, in a sequence of events, to detect early indicators. This does not prevent forecasting, however, since the models themselves are dynamic constructions simulating the evolution of ocean–atmosphere coupling based on an initial situation to which they apply the laws of fluid dynamics so as to be able to determine the subsequent states. To work, they do not need explicit causal chains.

In conclusion, the forecasting of ENSO several months in advance is realistic, but, because of the multiple interactions and its highly chaotic character, this predictability cannot go beyond one cycle. The forecasting of climate variations is still embryonic; it depends on improvements in the models. But these models are useless unless they are supplied with observational data from the real world. Although, for weather forecasting, there are operational atmosphere-observing systems, there is nothing comparable yet established for the ocean which is, nevertheless, the 'mastermind' of the system on the time scales of climatic variations. The experimental observation networks (TOGA) have proved their effectiveness, and remote-sensing satellites, such as Topex-Poseidon, have given proven performance. They must now be given an operational follow-up, without which we can never hope to achieve climate forecasting. It is true that to carry out *in situ* ocean observations on a permanent basis is no simple matter, unlike the atmosphere, for which the principal parameters required for forecasting (temperature, pressure, humidity, wind) are measured from stable platforms such as continents and islands. To penetrate the ocean and take the necessary measurements (temperature, salinity, currents), various kinds of platforms are needed: vessels, anchored instrument packages, automatic devices, instrumented drifting buoys at the surface and at depth, which can transmit their information by satellite, etc.

Moving to the operational phase implies means on a much bigger scale, given the cost of such installations, their replacement and their maintenance in a hostile environment. This is exactly what the World Meteorological Organization and the Intergovernmental Oceanographic Commission are aiming at by trying to convince their member states to set up a Global Ocean Observing System.

THE IRREGULARITIES OF EL NIÑO

There are two principal causes of irregularity that make of ENSO an oscillatory phenomenon resistant to standardization: the cycle of seasonal variations, and the atmospheric background noise. The classical version of ENSO traces the development of an El Niño event on the seasonal cycle. The delayed oscillator makes of it an oscillation mode specific to the equatorial Pacific the development of which, once triggered, is independent of the seasonal cycle. The TOGA observations show, however, that neither subjection to the seasonal cycle nor total independence of it corresponds to reality.

The ENSO process depends mainly on seasonal variations, which partly explain its chaotic behaviour and limit its predictability. In the foregoing analysis of ENSO, which is a multiannual phenomenon with a large amplitude, we neglected the seasonal variations because they are small in the equatorial zone. In the Pacific, the sea-surface temperature difference between a warm event and a cool event sometimes exceeds 6°C whereas it is usually less than 2°C over a 'normal' annual cycle. Even when masked in extreme periods (El Niño, La Niña), these variations are still present. In the northern-hemisphere winter and spring, the trade winds decrease along the equator, the intensity of equatorial upwelling weakens and the sea-surface temperature, which is the agent of ocean–atmosphere coupling and the motor of ENSO, increases. The opposite occurs in the opposite season. The seasonal cycle thus modulates ENSO and can even destabilize it and render its development chaotic.

The different reaction times of the two components of the climate system also play a role. On a meteorological time scale, the atmosphere is almost indifferent to the ocean. The opposite is not true, however, and the ocean, which controls variations on a climatic scale, is not indifferent to atmospheric fluctuations on the shorter meteorological time scale. These are the fluctuations called 'noise' in comparison to background noise which, speaking of a quiet meeting, the speaker should dominate. Over a sufficiently long time, the atmospheric noise seems to be random. It can, nevertheless, become sufficiently strong to be able to transmit to the ocean a signal that propagates as a wave and can trigger an El Niño event or disturb its development. This is the case of those westerly wind bursts that sometimes occur in the western equatorial Pacific and which have been proposed as a trigger of an El Niño, and, by recurring in 1996–97, have contributed to the originality of the event.

The effect of atmospheric noise itself also varies with the season, and the westerly wind bursts have a greater chance of triggering an ENSO at one point in the seasonal cycle than at another. These two sources of irregularity are therefore linked, which reinforces the non-linearity of the system and makes the risk of a chaotic development less predictable.

THE PAST AS A BASIS FOR LONG-TERM FORECASTING

The sea-surface temperature of the equatorial Pacific was particularly high in the 1980s and 1990s; correspondingly, the negative values of the southern oscillation index dominated during the same period. Calculated for the period 1950–88, the southern oscillation index anomalies were negative 36% of the time between 1950 and 1975 and 62% of the time thereafter. The magnitude of the 1982–83 and 1997–98 events and the abnormal duration of the 1991–95 event are undoubted signs of a change of regime already seen with the passage from the classical version typical of the period 1945–75 to the variability of the later situations. Is this the mark of a normal decadal variability or should we see in it the mark of an increasing greenhouse effect and of global warming that has been observed for a hundred years, with an increase in mean temperature of about 0.6°C? This warming paused momentarily between 1940 and 1975 only to take off again full blast and continuously since then (0.3°C since 1975). A coincidence? Some ascribe it to an increase in air temperature due to a warming of the ocean caused by the change of regime; others, in contrast, see this change in regime as a consequence of global change. The question remains unanswered; it can only be answered if we know the variability of ENSO on the decadal scale. Only a knowledge of the history of ENSO during previous centuries can clarify the matter. But this reconstitution is difficult, even if three kinds of information are available:

→ direct measurements of the characteristic parameters (atmospheric pressure, surface temperature, rainfall) or instrument measurements;
→ an indirect evaluation of these parameters based on environmental characteristics the evolution or growth of which depends on local climatic features: corals, trees, glaciers, which record this climatic variability;
→ documentary information left by those who have been witnesses, victims or even beneficiaries of these events.

We owe the first complete chronology to William Quinn, who used the information accumulated since the arrival of the Spanish in South America. With his collaborators, he published in 1992 a chronology of El Niño events since 1497, classified according to their strength: moderate, strong, very strong. They counted 124 events until 1987, or about one every four years. They extended these observations back to the Arab conquest in 622, by analysing the annual flooding of the Nile recorded in Cairo, starting from the idea that these floods were largely determined by the Blue Nile and the Atbara River draining the mountains of Ethiopia, themselves dependent on the monsoons, hence the southern oscillation.

By this reconstitution, El Niño became entwined in the more or less anecdotal events of the Spanish Conquest. For example, to go from Panama to Lima in Peru was not a simple matter, the winds and currents being adverse. It took several months or even a year. In any case, the duration of the trip certainly made a strong impression, as was witnessed by the report of a ship's captain who embarked at Paita in 1748 with his young wife for Callao: by the time he arrived, he was the father of a boy born during the voyage and who already knew how to read! It is likely that the demands of commerce rather than the weather explain the length of this voyage. Sometimes, however, the voyage was very quick: hardly twenty-six days from Panama to Lima in 1568 for Father Ruiz Portillo, who benefited from winds that were exceptionally from the north. Yet Quinn classified this particular year as one with a strong El Niño. The trace of ENSO could also be found in the fortunes of Spanish galleons which, after rendezvous with vessels coming from California and Peru, set sail from Acapulco for Manila, carried by the trade winds. The return trip was made at temperate latitudes in which westerly winds and currents prevailed. The records of the port of San Francisco reveal, nevertheless, that some voyages turned into catastrophes: a galleon coming from Lima could come up against overcast weather and a warm current coming from the north (El Niño) thus missing the rendezvous. Sometimes, sailors instead of having following trade winds came up against westerly winds; food became short and crews were decimated by scurvy. Alexander von Humboldt recounted the misadventures of a galleon captain, Don Josef Arosbide. Wishing to avoid ambush by British privateers, he tried to get to Callao from Manila by the direct route. Luck was with him and he did not have to sail against the trade winds, which had given way to weak but rather favourable winds; he made the voyage in ninety days, thanks to El Niño, that of 1791 which Quinn had considered to be a strong one. Unfortunately, Arosbide wanted to renew his exploit on the following voyage; after having vainly fought the trade winds, he was forced to sail northwards to take the usual route. Being short of food, he put into the port of San Blas where he died of fatigue and sadness.

No doubt more important for history is the following question: Did Francisco Pizarro have El Niño as an ally? This is what Quinn thought, judging from the report of Francisco Xeres, Pizarro's secretary on his 1531–32 expedition which took the conquistador onto the conquest of the Inca Empire, which led to the death of their chief, Atahualpa. Quinn noted first that Pizarro, leaving Panama in January 1531, reached San Mateo near the equator in thirteen days, whereas, on the previous voyage, two years had been necessary! Leaving San Miguel de Piura in northern Peru in September 1532, he arrived at Cajamarca where he laid his ambush for Atahualpa in November 1532 after having crossed without difficulty a country full of flooded rivers, which was unusual for that season. This romantic version has been challenged on the basis of a critical analysis of the original texts, as well as by a more exact reconstitution of the geographical references. Thus, Luc Ortlieb concluded, in 1999: no, there was no El Niño in 1531–32. If we have dwelt on this example, it is because it shows the difficulty of reconstructing a reliable chronology of climatic events based on writings open to the subjectivity of the author, in the first place, then to that of the reader. Ortlieb thus called into question the chronology of Quinn, excluding twenty-five events and incorporating seven new ones. It is difficult, under these circumstances, to carry out a reliable study of the variability of El Niño throughout history.

May we expect better results, using more rigorous scientific methods that allow an indirect evaluation of the climate parameters? Each growth ring in a tree represents a year, making dating easy, allowing us to go back in time from the periphery to the centre of the trunk. Some conifers in the western United States allow us to go back 300 or 400 years; we can even go back a thousand years using dead trees recovered by archaeologists. The annual growth is representative of the local climatic conditions; it increases with the temperature and the humidity. The thickness of the rings can therefore be analysed, providing information on annual climate variation during the tree's life. This dendrochronology is all the more sensitive when the climatic variations are most contrasted, as is the case in semi-arid zones. Such studies have been especially carried out in the south-western United States and in northern Mexico, both of which are teleconnected to ENSO.

Corals also allow such an approach, since their calcareous skeleton bears growth marks that allow us to go back in time several hundred years, sometimes with even a seasonal resolution. The skeleton's content of certain elements or the isotopic composition depends on the sea temperature, rainfall and productivity. Chemical and isotopic analyses of corals thus allow a reconstitution of the history of the environment in which they live. The islands and atolls of the Pacific, from the Galapagos to Indonesia, allow us to cover the equatorial Pacific Ocean, ENSO's domain. In this way, it has been confirmed that the variations in sea-surface temperature in the east were only partially representative of ENSO. In other words, the original El Niño, which occurs along the western coast of South America, is not entirely representative of ENSO! Significant anomalies of sea-surface temperature to the west and of the southern oscillation index do not always have a temperature signature along the American coasts.

Glaciers constitute a third kind of record. Each year, precipitation on them creates a new stratum whose properties (thickness, particle content, isotopic composition of the oxygen) are also characteristic of the climatic conditions at the time each layer was formed. Cores of some 200 metres, taken from Andean glaciers subject to the joint influence of the Pacific and the Atlantic, recount the climatic history of the last 1,500 years.

The ensemble of these recordings, plus the record of the Nile floods, provide us with an invaluable climate archive, but it is difficult to interpret it, if the aim is to use the data quantitatively to evaluate the variability of a climatic phenomenon such as El Niño. The difficulties are many and common to these different types of recordings. It is first necessary to be able to go from the archives to the climatic parameters (temperature, precipitation), which implies a solid relationship between the former and the latter. But this relationship, which translates the physical, chemical and biological processes, is neither simple nor unequivocal: several different climatic states can leave the same mark in the records. It is moreover necessary to carry out calibrations by comparing the data of the archives with instrument measurements of the contemporaneous climate parameters. So such calibration can only be carried out for the recent past (after 1850), for which such measurements exist. Moreover, each archive has an essentially local significance and we must be sure that it still represents well the regional climate with respect to the present. And finally, even supposing that the foregoing problems are resolved, a given archive tells its own story of the local climate variability and only reveals a particular but incomplete facet of ENSO, and the calibrations accept implicitly the hypothesis that the relationship between ENSO and the local climate is constant. We know this is not the case and that it is indeed one aspect of ENSO variability that we are seeking to understand. So there is a contradiction between the underlying hypotheses, which imply a certain constancy in the phenomena, and the objective, which is to determine ENSO's variability. It is for this reason that the conclusions of such studies are generally prudent. Thus, from a comparative analysis of the documentary historical information, the results from dendrochronology for Mexico and New Mexico, the analysis of cores taken from the Quelccaya glacier in the Peruvian Andes, Joel Michaelsen and Lonnie Thompson limited themselves to concluding that, since 1600, it is apparent that the variability of ENSO has hardly changed and that there is some evidence of periods of strong ENSO activity at the beginning of the 18th and 20th centuries and of a lesser activity at the beginning of the 19th century! To make any progress, the means must be found to synthesize the different records so as to recover the whole of ENSO's complexity. We are not yet in a position to explain the variability of ENSO and to derive the present trend from it – that is, the accentuation of El Niño episodes as a result of global warming – and even less to play the prophet by extrapolating its evolution.

7 El Niño on the witness stand

THE DIFFICULTY OF THE SOCIO-ECONOMIC APPROACH

Here we shall deal with the socio-economic consequences of the climate perturbations uncontestably associated with the year-to-year variability of the Pacific. To determine these consequences and their cost objectively, we must first draw up an overall balance sheet showing the costs and benefits corresponding to this climatic anomaly and define an 'economic anomaly' relative to a standard, as was done for the climatic parameters. In effect, the press gives too much attention to El Niño, attributing to it the catastrophic consequences of a good number of climatic anomalies; yet some 'positive economic anomalies' also arise when an El Niño occurs: mild climate and rainfall on the coasts of South America, which favours the vegetation (flowers bloom in the Atacama desert!), reduced cyclone activity in New Guinea, in the Philippines, in Japan, as well as in the area of the Gulf of Mexico, and a mild winter climate in North America, etc. This objective view is Utopian, for several reasons. No one is in a position to recount the beneficial effects of climate variations, or no one has any interest in recounting them, in the name of the common media view that good news is not always 'good' news. Also, the populations that suffer the damage do not benefit from the positive effects at the same time. Nor are the insurance companies, sources of much useful information, accountable for the catastrophes. So, we must not examine the consequences of climate variability only through the lens of disaster; beyond the media interest, there is one of economic reality. Even so, we must not confuse the cost of climate catastrophes with the real overall economic repercussions of climate variability.

Another difficulty arises in the economic approach. The evaluations, in effect, only take into account the replacement costs of what was lost or destroyed. They are therefore very much higher in industrialized countries than in developing countries. So they do not reflect the totality of the damage – far from it.

To be able to benefit from recent and reliable sources of information, to some extent we had also to take this 'natural catastrophe' approach. In this field, the risk and its cost depend on two factors: the phenomenon itself, which is more or less intense (the risk) and the vulnerability; that is the fragility of the infrastructure and the social and economic organizations where the catastrophe occurs. Thus, an earthquake of a given intensity may claim a few victims in California but kill thousands in, say, Cairo or in Armenia. When the catastrophe occurs, its cost is automatically imputed to the risk, while the vulnerability is forgotten. There are many examples: buildings in a zone of flooding carried away by the first flood, forestry and farming practices in Indonesia that lead to forest fires when the drought due to El Niño promotes them. We must not therefore conclude that the increase in the cost of natural catastrophes corresponds to an increase in their frequency or intensity.

EL NIÑO, THE PACIFIC AND ITS SURROUNDINGS

Certain planetary climatic events are systematically associated with El Niño (see Fig. G on p. 60) because they are directly linked to changes in the atmospheric circulation of the Pacific:

→ *drought and heat waves* in the western intertropical Pacific, which is often the site of strong rainfall – the 'marine continent' (Indonesia, Malaysia, New Guinea) and eastern Australia (the north of the State of Victoria and the cities of Melbourne and Sydney, the State of Queensland, the State of New South Wales, a major agricultural region, and a part of the Northern Territories);

→ *heat and high rainfall* lead to flooding of the coastal zone of South America which is usually dry and even arid – especially Peru and Ecuador, but also the Amazonian part of Bolivia, the Pacific coast of Colombia and northern Chile;

→ *cyclone activity* moving from the western side of the Pacific basin towards the triangle formed by Hawaii, Polynesia and the Cook Islands, which most of the time is free from them, and an increased frequency of cyclones originating in the Pacific but affecting Central America, and especially Mexico.

The marine continent, which is generally the site of ascendant air and the rainfall associated with it, undergoes a drought during El Niño. In February 1983, El Niño added to a period of drought affecting south-eastern Australia, bringing with it brush fires. A huge dust cloud spread over Melbourne: it arrived silently and a sudden darkness enveloped the city; some thought it was the end of the world. The atmosphere above was perfectly still and the terrifying dust cloud, 1 kilometre thick and containing half a million tons of dust, persisted for more than half an hour: Melbourne could no longer breathe. In 1997–98, nothing equivalent happened in this part of Australia. This drought also affected Indonesia, facilitating the advance of fires and affecting agriculture: in 1983, this drought came at a time when the country was just reaching self-sufficiency in food supplies and was even exporting rice. The Indonesian example illustrates very well the idea just mentioned, that a catastrophe is born of the conjunction of natural risks and human irresponsibility. The extreme drought that Indonesia experienced in the autumn of 1997 and the spring of 1998 explains the high occurrence of forest and peat fires which polluted the air over a part of South-East Asia, from Malaysia to the Philippines via Singapore and Indonesia. Even so, 2 million hectares of forest would not have burned if anarchy had not reigned in the logging industry. The major lumber companies bulldoze veritable 'boulevards' for the flames and burn out parcels of land to plant highly prized oil-palm trees.

On the other side of the Pacific, the particularly southward position of the ITCZ during El Niño accentuates the rainfall in Ecuador, in Peru and in neighbouring countries. The incessant rains sometimes attain 15 centimetres a day, bringing with them flooding and mud flows in Peru. On 15 February 1998, the River Piura, already a matter of concern in 1891 (see Chapter 2), overflowed, saturating the soil. 'Suddenly we were surrounded from all directions. Then my house just fell down completely,' declared Ipanaque Silva, a peasant from the village of Chato Chico. The flood water poured into the coastal desert of Sechura forming (see Fig. L on p. 64) a lake 160 kilometres by 40, with a depth of 3 metres, the second largest in Peru! These humid conditions also covered a part of South America, since Uruguay, Paraguay and Argentina were affected. In contrast, the whole of Central America underwent a period of drought.

Bolivia and Colombia, which are in an intermediate position between these climatic areas, experience mixed effects. In Bolivia, the highlands and valleys are affected by the drought during the first cereal harvest, whereas intense rainfall affects the Amazonian region. In Colombia, persistent heat and drought affect the north-west and the Andes, causing forest fires and the rationing of hydro-electric power. At the same time, some regions on the Pacific coast, and the inland provinces of Caquetá and Putumayo suffer heavy rains and flooding.

EL NIÑO, STAR OF THE MEDIA

The eruption of the Mexican volcano El Chichón in 1982, the El Niño episode of 1982–83, and the Chernobyl nuclear catastrophe in 1986 have, in a few years, shown that natural phenomena, or those caused by Man, know no frontiers. The fall-out from the Chernobyl radioactive cloud affected Europe; the ashes from the eruption of El Chichón, as those of Agung in Indonesia in 1963 and of Pinatubo in the Philippines in 1991, lowered the planet's mean temperature by a quarter of a degree by reducing the solar radiation of the oceans and the continents.

Recently, the attention of the media has focused on the El Niño/La Niña couple, so much so that politics have become involved. Thus, in Peru, the opponents of President Alberto Fujimori have reproached him for exploiting this disaster to gain a few popularity points (see Fig. L on p. 64). In California, the year 1997, the best since 1950 in terms of air quality, has been interpreted in different ways. For the *San Francisco Chronicle*, this was due to El Niño, which brought a cool and windy summer, thus avoiding fog formation. The view of the *Mercury News* differed; without denying the beneficial effect of El Niño, this Silicon Valley daily paper stressed the efforts of the state to reduce pollution, especially thanks to a new kind of fuel. The press headlines show that these events are associated with the notion of catastrophe, while the beneficial fall-out is constantly ignored:

- Climatologists Understand Better the '*Enfant Terrible*' of the Pacific – *Le Monde*
- Crazy Rains in the Tropics – *Le Nouvel Observateur*
- How Far will El Niño Go? – *VSD*
- The Calamitous Return of the 'Infant Jesus' – *Le Figaro*
- El Niño has Caused One of the Greatest Natural Catastrophes of the Century – *Le Monde*
- El Niño Stokes up the Fires of Asia – *Libération*
- How El Niño Inflames the Pacific – *Le Point*
- After El Niño, La Niña. The Terrible Twins of the Climate – *Sciences & Avenir*
- Contagious Diseases could Break out in the Pacific – *Le Quotidien du Médecin*
- El Niño, the Crazy Current that Throws the Climate out of Gear – *Science & Vie*
- El Niño, La Niña: Nature's Vicious Cycle – *National Geographic*

The Internet allowed us to follow the 1997–98 episode, day by day. The peak of the southern oscillation index, reached in February 1998, corresponded to a maximum of 120 articles a month on the Internet. The warm episode's development is underlined by the headlines which also show a premature announcement of the end of the El Niño:

- El Niño is Hatching in the Tropical Pacific – *ENN,* 18 June
- Prepare Yourselves for the El Niño Hell – *ABC News,* 14 October
- Is El Niño Dying Out? – *CNN,* 9 December
- El Niño will Go no Further – *MSNBC,* 14 January
- El Niño will Peak in the Coming Weeks – *The Irish Times,* 4 February
- It Continues to Advance – *ABC News,* 1 March
- El Niño is Lasting like a Mild Fever – *San Jose Mercury News,* 3 May
- Good-bye El Niño; Hello La Niña – *Los Angeles Times,* 27 June

ECONOMIC AND HUMAN CONSEQUENCES

In his book published in 1996 on the impact of El Niño on climate and society, Michael Glantz recounts his interviews with several scientists, especially César Caviedes, a son of Valparaiso. His first memory of El Niño was the catastrophic state of health of the thousands of birds deprived of food during the 1957 episode. Only a few people in the know linked this situation to the oceanic anomaly off the coast of Peru. For Caviedes, who became an oceanographer, El Niño now means: sea-surface temperature variations, rainfall anomalies, a decrease or an increase in the pressure field, southern oscillation index, teleconnection, etc. But it cannot be disconnected from its impact on the least-favoured human groups: the fisherfolk and aquaculturers of Peru, the rural populations in the *sertaos* or semi-arid zones of north-eastern Brazil, the llama-raisers of the Andean highlands, farmers and inhabitants along the Paraná River, herdsmen of sub-Saharan Africa, or the sheep ranchers of far-off Australia.

A FEW TEARS FOR THE PERUVIAN ANCHOVY?

Anchovy catches and El Niño

In terms of climatic impact, El Niños come and go, each one different from another, just like the anchovy catches (Fig. 7.1): their spectacular fall, after the 1972–73 episode, is the cause of the 'diabolization' of the El Niño phenomenon. In fact, the catches fell by more than 10 million tonnes in 1971, then about a quarter of the world fish catches, to 1.5 million only, in 1973. El Niño was designated the 'suspect', not altogether unreasonably, since it reduced or stopped the coastal upwelling, thus cutting the anchovies from their food supply. Things are not so simple, however, so a detour through the food web is necessary.

When human beings went from hunting and gathering to agriculture and breeding, they were only interested in plants or herbivorous animals. Any secondary steps in the trophic ladder represent, in effect, a loss of 90% of the organic matter: a tonne of forage yields only 100 kilograms of meat, etc. In the search for proteins from the sea, Man resorts mostly to fishing, aquaculture remaining marginal. The majority of the species caught are at the third or even fourth trophic level: for example, the tuna feeds on small fish which have eaten the zooplankton which, in its turn, fed on the plant plankton. In areas of upwelling, pelagic, phytoplankton-eating fishes are predominant. In contrast to sardines, which eat exclusively zooplankton, the anchovies, in their juvenile stages at least, consume phytoplankton directly, hence the high yield of the Peruvian trophic web.

In one year, in the Peruvian upwelling area, about 27 tonnes of anchovy are produced per square kilometre of sea surface. This efficacy has an adverse side: a high sensitivity of the trophic web to environmental fluctuations. When the supply of nitrates and phosphates ceases, the

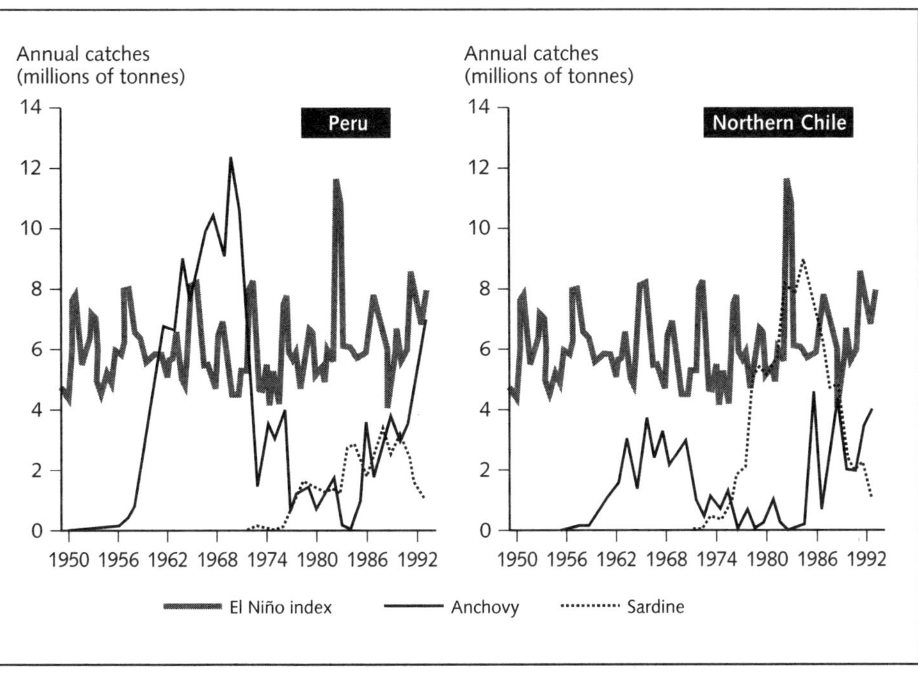

Fig. 7.1

Effect of El Niño on the anchovy and sardine fisheries off the coast of South America.

The death of millions of sea birds that feed on the anchovy provides the classical image of El Niño. The spectacular drop in catches after the 1972–73 event provided a basis for this scientifically sound idea, since, during a warm episode, the nutrient enrichment by upwelling is slowed down or even stopped, which affects the food web.

A comparison of the variation of the El Niño index and the catches of anchovy (*Engraulis ringens*) and sardine (*Sardinops sagax*) between 1950 and 1992 shows that this idea should be adjusted, at least. In effect, the fishery also depends on long-term fluctuations in the environment, on the fishing effort and on the spectacular improvement in fish-catching techniques.

The anchovy, the juveniles of which consume phytoplankton directly, is favoured by the activation of upwelling, whereas the recruitment to the sardine stock, which feeds exclusively on zooplankton, is optimal when the water is less cool. Even when El Niño has an adverse effect on the fishery, it favours an increase in biodiversity by bringing with it new tropical species. There is a growing interest in sardine catches in Peru, and especially in Chile where they reached 8 million tonnes a year in the mid-1980s.

primary production, then that of the partially phytoplankton-eating fishes, like the anchovy, fall considerably. The latter then leaves room for other species, such as the sardine, hence for a lower yield. The ecosystem of the Peruvian coast then only produces 1 tonne of fish per square kilometre. From some 4 million tonnes between 1974 and 1976, catches fell to 1.5 million, and this, right up to 1985. After this time, growth was resumed and the catches have since reached a level comparable with those of the pre-1972 period (nearly 10 million tonnes in 1994) in spite of the warm episodes of 1986–87 and 1992–95. Even the El Niño of the century (1982–83), which after all had reduced the catches to their minimum in 1984, to 23,000 tonnes only, could not prevent a rise in the fish catches to 3.5 million tonnes two years later.

The fishery and long-term fluctuations

Is not El Niño just an epiphenomenon that grafts itself onto a longer-term variation reinforcing or limiting its effects? We come back to the same question as before regarding the interactions between the different scales of variability and, particularly, the influence of decadal variations on the strange variability of El Niño. This relatively constant phenomenon between 1945 and the mid-1970s (the classical version) afterwards becomes much more chaotic. It is tempting to link the stagnation in the anchovy catches between the mid-1970s and the early 1990s to this change in the ENSO regime during the same period in which there was a predominance of negative anomalies in the southern oscillation index (see Chapter 6). This is what Eleuterio Yañez and his collaborators are doing; they have observed, over the same period, a slight temperature increase along the coasts of Peru and northern Chile, and a strong increase in sardine catches. From a level of less than 100,000 tonnes annually, up to 1975, the Peruvian sardine catches increased to more than 1 million tonnes as from 1978, exceeding 3 million tonnes between 1988 and 1991. It is notable that, in 1984, following the 1982–83 El Niño, when anchovy catches reached their minimum (25,000 tonnes!), the sardine catches doubled, reaching nearly 3 million tonnes!

It is therefore logical to speak of competition between the anchovy, favoured by upwelling of cool water, and species, principally the sardine, in which recruitment is optimal when the water is warmer and the upwelling less intense. For the industrial fishery, the result is obviously not the same, since the 'cool' periods, in which the anchovy are predominant, are much more profitable than those in which the sardine and other species higher in the food chain are more abundant. Thus, the coastal upwelling ecosystem off Peru and northern Chile depends considerably on climate fluctuations on a decadal scale. This must be kept in mind if the real impact of El Niño on the fishery is to be fully understood.

During a cool period the ecosystem is more profitable because the trophic chain is short; on the other hand, it is also much more fragile, because it is practically monospecific. When El Niño occurs, the anchovies try to find a more favourable biotope, thus fleeing the incoming warm water or swimming to greater depths towards cooler water. In any case, they become inaccessible to the fishing gear, and the catches fall. Moreover, the cessation of the upwelling creates tricky conditions for the development of the larvae and juveniles; recruitment is greatly reduced and so, temporarily, is the stock. The economic consequences are obviously not negligible, but the situation returns to normal rather rapidly. Miguel Carranza has, moreover, noted that, since 1998, in the post-El Niño period, there has been a remarkable reappearance of juvenile anchovies, which shows the rapid speed of the biological system's response.

During a warm period, the fishery is less profitable, but the situation ensures a greater species diversity of non-phytoplanktivorous pelagic and demersal species which, because they depend less directly on fluctuations in the upwelling and in the primary production, can more easily dampen the disturbance that El Niño brings to the ecosystem. This was proven by the low sensitivity of the Peruvian sardine catches to so marked an event as the 1982–83 El Niño.

Nature does not win every time: to a productive and profitable system there corresponds a higher risk; and to a more diversified and less profitable system, there corresponds a much lower risk. Even if that does not satisfy current criteria of profitability, it is possible to say that El Niño compensates for the temporary drop in productivity by an increase in biological diversity. It is this renewed diversity that allows the fishermen to discover and to fish new tropical species which the sailors of Paita have welcomed by giving the phenomenon the now well-known name of El Niño. In any case, El Niño, which cannot be held responsible for the lasting drop in catches after 1973, is probably indispensable to the good health of the ecosystem, which would be hard put, in a context of intensive fishing, to maintain itself under the quasi-monospecific conditions of the cool episodes if it did not, from time to time, get a little increase in diversity from El Niño.

ENSO AND THE WATER CYCLE

The southern oscillation leads, above all, to important changes in the water cycle. Yet water, whether in short supply or causing floods, is the main scourge, either directly or through the diseases associated with it.

Using one hundred years of data (1890–1989), the IRI (International Research Institute for Climate Prediction) has analysed the rainfall anomalies during the twenty warmest years (El Niño) and the twenty coolest years (La Niña) and compared them with twenty 'normal' years. Some ten sites worldwide were inventoried; some of these do not, in our view, incontestably fall with ENSO's area of influence, so we shall limit ourselves to a few examples clearly connected with this phenomenon.

The Indonesian region is vast, covering more than 7 million square kilometres, from 10°S to 5°N and from 100°E to 150°E. The results leave no doubt: nine times out of ten, the rainfall between June and November is deficient in an El Niño year. In four cases out of twenty, this deficit reaches or exceeds 6 centimetres per month; in seven other cases, it exceeds 3 centimetres. The data are even more pronounced for La Niña, since, in all cases, the rainfall is in excess; in one case out of two, the excess is greater than 3 centimetres per month. In the tropics, Africa is mainly affected by the drought in West Africa, in the Sahel and in southern Africa. Regarding the Sahel, the 1982–83 El Niño and, especially, the 1972–73 episode aggravated a drought that had been recurring since 1968. In 1982–83, the drought particularly hit South Africa and Zimbabwe, both cereal exporters: South Africa was obliged to import 1.5 million tonnes of wheat from the United States, and Zimbabwe called on international aid to avoid a famine. These two countries therefore prefer a La Niña year!

As a general rule, the areas hit by drought during El Niño experience very heavy rainfall during La Niña; those that experience mild winters during warm episodes have harsh conditions during La Niña, and so on. La Niña, which is a cool episode, counterbalances the effects of El Niño on the energy cycle. La Niña leads to an increase in rainfall in South-East Asia, especially during the south-west monsoon, in northern and north-eastern Australia, in southern Africa, in northern South America, including north-eastern Brazil, in Central America, and in the Hawaiian Islands. In contrast, La Niña causes a drier than normal climate in the islands of the central equatorial Pacific, in eastern central Africa, during the little rainy season, along the Gulf of Mexico, in the south-western United States and to the north of Mexico, as well as in certain regions of southern South America.

Americans, who are particularly concerned by the devastating effects of hurricanes, invest a great deal in forecasting them. This is especially true of those originating in the Atlantic, which are the most frequent. They have a particular interest in La Niña, which favours hurricane activity in this ocean. Thus, in 1995, eleven hurricanes crossed the coast into the United States; they included Hurricane Marilyn, which devastated the Virgin Islands, and Hurricane Opal, which devastated Florida. Hurricane Linda, which approached Mexico in September 1997, with winds easily exceeding 300 kilometres per hour, was one of the strongest ever in terms of energy content. NOAA data covering 98 years, including 23 El Niño years and 15 La Niña years, are eloquent: on average, 1.04 hurricanes hit the United States during a warm (El Niño) year, 1.61 in a neutral year, and 2.23 in a La Niña year.

Perturbations in the water cycle also cause health problems, mainly respiratory affections, in human beings and animals; medical consultations increased tenfold in Malaysia following the huge forest fires in Indonesia. Usually, the diseases are insect-borne infectious diseases (malaria, Rift Valley fever) or diarrhoeic (cholera and shigellosis), which break out because the larval development of the vectors is favoured in flooded areas. Certain data of the World Health Organization (WHO) illustrate our reservations on this evaluation, however. For example, WHO indicated that, during the 1997–98 episode, the number of malaria cases reached very high values in Bolivia, Colombia, Ecuador, Peru and Venezuela, and increased four or five times in Pakistan and Sri Lanka. Yet, in the latter, which are in a monsoon region, an El Niño year is generally characterized by a decrease in rainfall, hence a decrease in malaria cases. In 1997–98, El Niño was blamed for the fact that it did not, a priori, bring about a weakening of the monsoon as it should have done!

Rift Valley fever can also be cited; it affects cattle and human beings in the Horn of Africa, and any flooding leads to the eclosion of the *Aedes* mosquito which is infected by the virus. En 1997–98, the loss of cattle was very high, and 90,000 people were infected, with more than 200 dead in north-eastern Kenya and in southern Somalia. Epidemics of cholera and other diarrhoeic illnesses are as bad during flooding as during drought, since they are very responsive to water contamination. Peru counted nearly 17,000 cases, including 150 deaths, during the 1997–98 El Niño, but the count was much higher in the Horn of Africa, with 40,000 cases and more than 2,000 deaths in Tanzania in 1997. The other countries of this region were also affected: 17,200 cases (555 deaths) in Kenya, 6,814 (252) in Somalia. For the first quarter of 1988, Kenya had already recorded more than 10,000 cases, with 507 deaths, and Uganda, 110,335 cases, including 525 deaths.

SOME SOCIO-ECONOMIC DATA ON THE 1982–83 AND THE 1997–98 EPISODES

The two warm episodes, of 1982–83 and of 1997–98, led to intensive study of their socio-economic impacts, carried out, notably, by international bodies linked to UNESCO. The insurance companies also played an important role. But this did not lead to a very objective scientific assessment: first of all, because the positive consequences were neglected, then, because the assessments include events whose relationship with ENSO seems doubtful or at least insufficiently proven.

The 1982–83 event, sometimes qualified as being an 'abnormal anomaly', because the ascending side of the atmospheric circulation cell was displaced 8,000 kilometres, claimed 2,000 victims and the damage it caused was estimated at some $10 billion. In effect, El Niño caused cyclones in Polynesia and in Hawaii. It caused flooding in Bolivia, Ecuador, northern Peru, Cuba and the United States (Gulf of Mexico). Finally, it was the cause of drought episodes that led to a drop in harvest (maize in Zimbabwe), and of huge fires that affected South Africa, southern India, Sri Lanka, the Philippines, Indonesia, Australia, southern Peru, western Bolivia, Mexico and Central America.

The warm episode of 1997–98 caused more than 2,000 deaths and caused damage estimated at some $25 billion; it is not certain that this cost estimate can be compared to that of 1982–83 because the basis of calculation was different. To give an idea of its importance, we can point out the fact that the average cost due to the impact of climatic risks is about $40 billion per year, although some consider these figures underestimates.

To put the cost due to the 1997–98 El Niño into perspective, we can indicate that the exceptional flooding in China, which was not taken into account in the evaluation because the connection with the southern oscillation remains to be proved, alone caused more than $30 billion worth of damage.

The payments made by the insurance companies constitute another way to approach natural catastrophes. All records were beaten in 1998: 350 events, more than 22,000 victims, nearly 5 million homeless, about $25 billion worth of damage. Moreover, the insurers react more quickly than the scientists! For the latter, this El Niño was less devastating than that of 1982–83, which therefore remains the event of the century, and an increase in the frequency of destructive phenomena related to global warming is probable.

A word on natural catastrophes. Geological phenomena cause 40% of the deaths (earthquakes: 34%, volcanic eruptions: 5%, landslides: 0.08%, tsunamis: 0.001%). The remainder is attributable to climatic caprices, especially to cyclones, which are responsible for 60% of the victim. The other meteorological phenomena – flooding unrelated to cyclones, thunder- and rain-storms, cold spells or heat waves – have only a minor impact. This term, natural catastrophe, is ambiguous. First of all, because human activities help to trigger some phenomena that are qualified as 'natural'; global warming is a good example. Secondly, and especially, because the gravity of the social impact depends on the social vulnerability. Industrialized countries and developing countries are not equal; life 'has no value' in poor countries, as the two first lines of Table 7.1 on the 1997–98 episode show. The cost of damage was estimated at about $5 billion in the United States and at less than $100 million in Africa. In contrast, more than 13,000 victims were counted in Africa and fewer than 600 in the United States.

Table 7.1
Socio-economic consequences of the 1997–98 El Niño

	Africa	Asia	Indonesia and Australia	North America	Central and South America
Cost[1]	0.1	3.3	4.45	5.5	15.0
Mortality	13 325	5 648	1 316	559	858
Morbidity	107 301	124 647	52 209	?	25 696
Displaced	1 357 000	2 555 000	1 443 000	410 000	363 000
Area affected[2]	190 755	1 544 701	2 812 480	12 315 600	5 640 876

1. In billions of dollars.
2. In hectares.

Of the ten most costly disasters since 1995, nine occurred in industrialized countries of North America and Europe: United States (Hurricane Fran in September 1996, blizzards in the north-west in January–February 1996 and in January 1998, tornadoes in the Midwest in May 1998), the United Kingdom (flooding in April 1998, storms in December 1997, snow and flooding in January 1998), Canada (icy rain in January 1998) and central Europe (flooding in July–August 1997). The only other country on this list was China, with its flooding in the summer of 1998.

8 The outlook

Throughout the parallel story of El Niño and the southern oscillation we have encountered climate forecasting; the quality of this forecasting has steadily improved, in spite of setbacks or, rather, partly because of them. The more the science of climate progresses, the more demanding its 'clientele' becomes! The same goes for meteorological forecasting; is not the complaint made by French farmers against the weather forecasters, on the grounds that they were not able to forecast the force of a storm, also a kind of tribute to the habitual know-how of Météo France?

The challenge of making a reliable forecast several months in advance for each region of the Earth is a difficult one, because the Earth's climate is ill-fitted to the separate study of its components: the atmosphere, the oceans, the continents and the rivers, the cryosphere and the biosphere, including human beings. Each component has its own physical, chemical and biological characteristics, as well as its own dynamics; but the agents of the climate system disregard frontiers.

Water, the main agent of energy transfer, is provided to the atmosphere by evaporation from the surface of the ocean and the land, and by evapotranspiration from the vegetation. Its condensation gives rise to rainfall that supplies the oceans, the continents, the living world, glaciers and polar ice-caps. Carbon dioxide, which allows the creation of living matter by way of photosynthesis, is also a greenhouse gas abundantly produced by human activities, which can cause global warming. It is exchanged by physical, chemical and biological processes amongst all the environmental media, where it occurs in the dissolved, gaseous, particulate, mineral or organic form.

Climate forecasting implies not only knowledge of the dynamics of each of the climate compartments, but also of their interactions; what is known as coupling. The ideal model for climate forecasting should therefore couple the different compartments, without forgetting the activities of human beings, an important factor in climate change. Nor should the Sun be forgotten, the initial source of energy, the intensity of which varies, nor the Earth's interactions with the other components of the solar system that affect its trajectory.

The study of complex objects depends on the tools that science has available. Human beings, in their conception of the world, and especially of the planet that houses them, are slaves to their senses and to their extensions by the tools of observation that they have created. The scientist therefore gives preference to phenomena that occur on time and space scales accessible to the available means, in an approach often qualified, with scorn, as 'reductionist'. With due respect to the supporters of complexity, it is nevertheless to observation and the analysis of simple mechanisms that we must turn to identify complex systems as objects of science and not of mythology. But these means of observation, which impose on scientists their own time and space scales, also put blinkers over their eyes, and they are sometimes inclined to attribute whatever goes against their concepts to error of measurement. The transition from the Paita fishermen's El Niño to the ocean climatologists' ENSO, then to the whole of the planetary climate system on all its scales of variability, is an example of this growing 'complexification' which has accompanied more than guided advances in techniques of direct observation, up to the space revolution.

The story of ENSO, at each stage, invariably brings us back to the following question: Is it possible to forecast the climate system? To want to reply to this question inevitably leads to this second question: Do we have the means of knowing the climate sufficiently well to simulate its evolution on various time and space scales in order to allow an operational forecasting system to be set up? To answer these questions, researchers construct models that, before becoming prediction tools, are a means of experimentation on the functioning of the climate system. Based on the 'laws' that govern the dynamics of the climate compartments and their interchanges, they allow us, starting with a given state of the system, to simulate its future state. With such a tool, the researcher can test the behaviour of the ocean–atmosphere system, after having introduced, for example, a modification of the sea-surface temperature in the equatorial Pacific, or of the southern oscillation index, or even develop climate scenarios in the event of a doubling of the concentration of greenhouse gases between now and the end of the twenty-first century, etc.

For this game to be productive, it is necessary to know the laws used to develop it. The term 'law' only indicates the mathematical relationships between the parameters, derived from laboratory measurements or *in situ* observations which improve as the relevant techniques progress. The simulations carried out using models also allow us to test the impact of these improvements on the quality of the model. The only real validation of such simulations is by confrontation with reality, which implies observation networks adapted to the spatio-temporal scales of interest. So models and observation systems are inseparable.

Climate models coupling the atmosphere and the ocean already exist. Their extension to other components of the system, such as the cryosphere and the land surfaces, depends principally on the calculating capacity of computers; since this capacity is growing constantly, this is not, in the short term, an obstacle. Regarding the observation networks, it is 'simply' a problem of cost! Satellites have revolutionized the observation of the Earth and, in particular, its climate system. Who has not felt the uniqueness of this system when looking at images obtained from meteorological satellites and the animations of cloud systems and complicated eddies, forming and breaking up as they move around the Earth? Using instruments mounted on the satellites, we can measure over the whole planet the main climatic factors: air temperature and humidity, wind speed and direction, sea-surface temperature, changes in ocean currents (by satellite altimetry), soil temperature and humidity, marine and terrestrial primary production, area and reflective power of sea ice, etc. And the precise localization and transmission of data by satellite allows us to increase considerably, all over the world, the number of automatic stations for the measurement of climatic factors: fixed stations or mobile stations (on board ships or aircraft) or even drifting freely in ocean currents, on the surface or at depth.

The systems have been tried out since the 1980s under the World

Climate Research Programme, based on an analytical approach to climate. Some operations were concerned only with one of the compartments of the system (ocean circulation under WOCE – World Ocean Circulation Experiment) or even for a time scale of a given variability (TOGA, mentioned previously). The possible modelling of the climate system and the new means of observation allow the World Climate Research Programme to advance to a higher level of complexity by launching a Study of Climate Variability (CLIVAR) on a decadal scale, linking oceans, atmosphere, land and cryosphere to arrive at operational forecasting. It is certainly not yet a unified climate-observing system since the subtle interactions of the biosphere and of the climate have been left to other programmes. Nevertheless, it is an important step forwards, since this programme links research with concern for operational forecasting. This is how meteorology has advanced to reach seven-day forecasting and now to experiment with two-week forecasting. The key is the establishment of a long-term observation system, specifically for the ocean: satellite remote sensing, drifting buoys, repeated oceanographic cruises, etc.

What is left of ENSO and its episodic fits called El Niño and La Niña in this globalization of the climate system? First of all, a better knowledge of the scale of climatic variability in phase with the rhythm of human activities, and this is crucial. To give most attention to this time scale is also part of the scientific logic. In effect, the more we aim at long projections, the longer the time-series of observations we need; however interesting they may be, palaeoclimatic data (growth rings in trees, glaciers, sediments, corals) which allow us to reconstitute past climates cannot make up for the lack of real, long-term time-series of observations. The ENSO time scale allows the scientist to test rapidly the quality of predictions derived from models and thus to improve them. Finally, the amplitude of the climatic perturbations associated with ENSO means that, even within the complexity of the climate system, it affects human activities too directly to be forgotten: El Niño and La Niña will continue to be talked about for a long time to come!

Further reading

Anon. 1998. Les humeurs de l'océan. *Pour la science* (Paris), October 1998.

——. 1999. El Niño, La Niña – Nature's Vicious Cycle. *National Geographic* (Washington, D.C.), March.

Arntz, W. E.; Fahrbach, E. 1991*a*. *El Niño: experimento climático de la naturaleza.* Mexico City, Fondo de Cultura Económica. 312 p. (Also published in German, see below.)

——. 1991*b*. *El Niño – Klimaexperiment der Natur: Physikalische Ursachen und biologische Folgen.* Basel, Birkhaüser-Verlag. 264 p. (Also published in Spanish, see above.)

Bigg, G. R. 1998. *The Oceans and Climate.* Cambridge, Cambridge University Press.

Chapel, A. et al. 1996. *Océans et atmosphère.* Paris, Hachette.

Fierro, A. 1991. *Histoire de la météorologie.* Paris, Denöel.

Gautier, Y. 1995. *Catastrophes naturelles.* Paris, Presses Pocket.

Glantz, M. 1996. *Currents of Change: El Niño's Impact on Climate and Society.* Cambridge, Cambridge University Press.

Glantz, M.; Katz, R.; Nicholls, N. 1987. *The Societal Impacts Associated with the 1982–1983 Worldwide Climate Anomalies.* Boulder, Colo., NCAR–UNEP.

Graedel, T. E.; Crutzen, P. J. 1997. *Atmosphere, Climate, and Change.* New York, Scientific American Library.

Jacques, G. 1996. *Le cycle de l'eau.* Paris, Hachette.

Kandel, R. 1998. *Les eaux du ciel.* Paris, Hachette.

The Open University. 1989. *Ocean Circulation.* Oxford, Pergamon Press.

Philander, G. 1983. ENSO Phenomena. *Nature* (London), No. 302.

——. 1986. Predictability of El Niño. *Nature* (London), No. 321.

——. 1990. *El Niño, La Niña and the Southern Oscillation.* San Diego, Calif., Academic Press.

Rasmussen, E. 1984. The Ocean/Atmosphere Connection. *Oceanus* (Woods Hole, Mass.), No. 27.

Rasmussen, E.; Wallace, J. M. 1983. Meteorological Aspects of ENSO. *Science* (Washington, D.C.), No. 222.

Rochas, M.; Javelle, J.-P. 1993. *La météorologie, la prévision numérique du temps et du climat.* Paris, Syros.

Sadourny, R. 1994. *Le climat de la Terre.* Paris, Flammarion.

Voituriez, B. 1992. *Les climats de la Terre.* Paris, Presses Pocket.

World Meteorological Organization. 1999. The 1997–1998 El Niño Event: A Scientific and Technical Retrospective. Geneva, WMO. 96 p. (doc. WMO No. 905).

Wyrtki, K. 1978. Predicting and Observing El Niño. Science (Washington, D.C.), No. 191.

Glossary

Anticyclone

Area of high atmospheric pressure.

Anticyclonic

Qualifies a horizontal eddy motion of the atmosphere or the ocean in a clockwise direction in the northern hemisphere (and in an anti-clockwise direction in the southern hemisphere) around zones of high atmospheric pressure.

Atmosphere

Gaseous envelope around planets. The Earth's atmosphere consists of nitrogen (77%), oxygen (21%), argon (1%), water vapour, carbon dioxide and other gases in very low concentrations. Meteorological and climatic phenomena occur in the lower layer of the atmosphere: the troposphere (from the ground up to 7 kilometres at the poles and up to 20 kilometres at the equator); and in the stratosphere (up to about 50 kilometres above ground level).

Autotroph

An organism that elaborates its own living matter exclusively out of inorganic elements, by chemosynthesis or photosynthesis.

Benthic

Living on the sea floor.

Biocenose

A community of plants and animals that occupy a given surface or volume; that is, a **biotope**. So: biocenose + biotope = **ecosystem**.

Biodiversity *see* **Diversity**

Biomass

Amount of living matter present at a given time in a given area or volume.

Biotope

Surface (or volume) with uniform physical and chemical characteristics occupied by a species or, more generally, by a specific community (**biocenose**).

Carnivore

An animal that depends principally or exclusively on other animals for its food.

Convection

Vertical movement of a mass of air or of water resulting from instability in the density field usually of thermal origin. In the heated fluid the warmer parts rise and the cooler ones sink forming a circulation cell. Convection may lead to **convergence** and **divergence**.

Convergence and **Divergence**

Convergence is a zone of confluence and divergence is a zone of separation of two air masses or two water masses. Since they occur in the horizontal plane, convergences and divergences cause compensatory vertical movements. At the ocean–atmosphere interface, a convergence of winds (the **ITCZ**, for example) causes an ascent of the air at the base of the circulation cell. A convergence of surface-water masses, by contrast, causes a descent of water since it is at the top of the oceanic circulation cell.

Coral

Benthic anthozoans existing as individuals or colonies that deposit calcium carbonate in their skeletons. Under certain conditions, these animals create coral reefs together with calcareous algae.

Coriolis force

A force that displaces any moving body, as a result of the Earth's rotation about the poles. The force operates perpendicularly to the velocity of the moving body, to the right in the northern hemisphere and to the left in the southern hemisphere.

Cromwell Current or **Equatorial Undercurrent**

A current flowing eastwards along the equator within the layer of maximum thermal gradient (**thermocline**) but in the opposite direction to the **South Equatorial Current** at the surface.

Cyclone

A region of low pressure in a limited area of the tropics in which the air is drawn into a whirlwind with wind speeds exceeding 200 kilometres per hour (see **cyclonic**).

Cyclonic

Qualifies a horizontal eddy movement in an anti-clockwise direction in the northern hemisphere (and in a clockwise direction in the southern hemisphere) around areas of low atmospheric pressure.

Delayed oscillator

Theory proposed to explain the development of El Niño/La Niña episodes on the basis of interferences, along the equator, between oceanic **Kelvin waves** and **Rossby waves**.

Dendrochronology

A method of dating based on growth rings in trees. Variations in the thickness of the rings allow past climatic changes to be reconstructed.

Divergence *see* **Convergence**

Diversity

Specific diversity corresponds to the number of species found in an environmental subdivision or to an index expressing the type of distribution of individuals within the species. There is also biotic diversity or, now, biodiversity to express the variety of life styles and environments.

Downwelling

Sinking of surface water, especially in oceanic areas of **convergence**.

Ecosystem

A functional unit consisting of organisms (**biocenose**) and environmental factors (**biotope**) of a specific area or volume.

El Niño

Initially, a warm marine surface current flowing southwards and developing off the coast of western South America (Ecuador–Peru). Now, this term designates the warm episode of **ENSO** characterized by a markedly negative **southern oscillation index** and abnormally warm ocean temperature at the equator and in the eastern Pacific, as well as a weakening of the **Walker cell**.

ENSO = El Niño – Southern Oscillation

An oscillation of the atmospheric pressure field between the area of high pressure in the central Pacific and the low-pressure area in the Indo-Pacific region. This oscillation is coupled to the variations in the sea-surface temperature of the Pacific Ocean.

Equatorial

Relating to the equator. Region near the equator.

Equatorial Counter-current

Ocean current flowing eastwards between the **North and South Equatorial Currents** along the meteorological equator.

Food web

The ensemble of organisms of an **ecosystem**, from the primary producers to the highest levels of the food chain. The flux of material and energy between these various levels, from the **autotrophs** to the herbivores and the various levels of **carnivores**.

Geostrophic

Qualifies the approximation by which the horizontal pressure gradient balances the **Coriolis force**. The geostrophic method is used to calculate currents, based on this approximation.

Glacial (episode, period)

A period during which the high and intermediate latitudes are covered with continental glaciers. The Pleistocene Epoch, the most recent division of the Quaternary Period, includes the last Ice Age. Ice Ages last about 100,000 years and have recurred every 120,000 years for at least the last million years.

Greenhouse effect

Warming of the Earth's atmosphere as a result of absorption, by atmospheric components such as water vapour and carbon dioxide, of infra-red radiation emitted by the Earth. This natural effect ensures that the Earth has an average temperature of 15°C. Human activities accentuate the greenhouse effect and can therefore modify the climate.

Greenhouse gas

Any gas that, because of its strong capacity to absorb infra-red radiation, contributes significantly to the **greenhouse effect** (water vapour, carbon dioxide, etc.). Human activities produce such gases, especially carbon dioxide, methane, chlorofluorocarbons, and thus increase the **greenhouse effect**.

Guano

Consisting of sea-bird droppings, it is a powerful nitrogenous fertilizer.

Hadley cell (or **circulation)**

Meridional atmospheric circulation marked by the ascent of warm and humid air (**convection**) over the **ITCZ** and its descent over areas of high subtropical atmospheric pressure where the major deserts are found.

Humboldt Current

Name given to the ocean current that flows from south to north along the west coast of South America. Also known as the Peru Current.

Interglacial (episode, period)

Short periods in the Quaternary, lasting from 15,000 to 25,000 years, separating two Ice Ages, and during which the Earth experiences a warmer than average climate.

Intertropical

Covering the zone between the two tropics (of Cancer and of Capricorn), hence including the equator, hence the term's ambiguity.

ITCZ = Intertropical Convergence Zone = meteorological equator

Zone towards which the trade winds in both hemispheres blow. On average, it is situated 5 degrees to the north of the geographical equator. Its position varies with the season, being displaced northwards in the northern-hemisphere summer. It also corresponds, in a north–south sense, to a maximum in the sea-surface temperature.

Jet stream

Very strong currents in the upper layers of the troposphere.

Kelvin waves

Oceanic waves generated by atmospheric perturbations and propagating along the equator from west to east.

La Niña

Cool **ENSO** episode during which the **southern oscillation index** is strongly positive. At the same time, in parallel, an activation of the **Walker cell** in the Pacific is observed, as is a marked cooling of the surface water in the eastern Pacific and along the equator, corresponding to the activation of coastal **upwelling** and of equatorial **divergence**.

Meteorological equator *see* **Intertropical Convergence Zone Model**

Simulation of a natural phenomenon. It can be physical (a reduced-scale model) or mathematical, using equations to represent the phenomena. The models used in climatology and in oceanography are mathematical models that are resolved using numerical simulations.

Momentum

A physical quantity (mass times velocity) that is conserved in kinetic-energy exchange between interacting bodies. The entrainment of ocean currents by the wind corresponds to the transfer of an amount of momentum between the atmosphere and the ocean.

Monsoon

The name given to the seasonal winds (and derived from the Arabic word for season: '*mausim*'). The term was originally applied to winds over the Arabian Sea blowing from the south-east in summer and from the north-east in winter.

North Equatorial Current

Current flowing westwards under the stress of the trade winds in the northern hemisphere.

Nutrients

Chemical elements necessary for aquatic photosynthesis. The term is often restricted to elements whose concentration in the aquatic medium, when very low, may limit photosynthesis. This means the term is, in this medium, effectively synonymous with the inorganic forms of nitrogen, phosphorus and silicon.

Overfishing

Excessive fishing such that the juveniles are no longer able to rebuild the stocks being overfished.

Pelagic

Qualifies the open-water medium and the life it contains, the pelagos, which includes the **plankton** and the ensemble of swimming organisms, or nekton (cephalopods, fishes, marine mammals, etc.).

Phytophage

Animal that consumes plants. Especially used for aquatic animals that consume **phytoplankton** (**zooplankton**, fishes and even mammals, such as the baleen whale).

Phytoplankton

Plant plankton formed of microscopic photosynthetic organisms with a size between less than a micrometre and a millimetre.

Phytoplankton bloom

Phytoplankton, which consists of microscopic plants, multiplies very rapidly under favourable conditions. It has the potential to double its **biomass** each day, reaching concentrations of several million cells per litre of sea water; it can then give a strong colour to the water.

Plankton

Organisms living in open water (the **pelagic** medium) whose capacity for independent movement relative to the water itself is very low.

Plate tectonics

Theory according to which the lithosphere (the Earth's crust) is broken up into plates that move relative to each other under the force of convection currents in the mantle. Most seismic and volcanic activity occurs in the boundaries between the plates.

Polar ice sheets

Major polar glaciers that presently cover Greenland and the Antarctic. During an Ice Age, in the northern hemisphere, the polar ice sheets also cover Canada, northern North America and Eurasia.

Primary production

Amount of living matter produced by **autotrophs** (primary producers) per unit of sea surface (or volume) and of time.

Recruitment

In a fishery, this term designates the phase and the quantity of fish reaching a stage at which their individual size makes them accessible to the fishery.

Remote sensing

Presently used for methods requiring sensors mounted on board aircraft or, more usually, satellites (satellite remote sensing).

Rossby waves

Waves propagating in the atmosphere or in the ocean from east to west. Their speed depends on the stratification of the medium and decreases as the latitude increases.

Sea-surface topography

Mapping of the sea level relative to a reference surface, or geoid. The **Topex-Poseidon** altimeter is used to elaborate such maps.

South Equatorial Current

Current flowing westwards under the stress of the trade winds in the southern hemisphere.

Southern oscillation *see* **ENSO**

Southern oscillation index (SOI)

An index describing the variation of the southern oscillation (see **ENSO**). It is the difference in atmospheric pressure at sea level between Tahiti and Darwin (Australia).

Subtropical

Around 30° latitude, just to the north of the Tropic of Cancer and just to the south of the Tropic of Capricorn.

Thermocline

The suffix *-cline* signifies a layer in which the physical or chemical properties show a strong gradient. The thermocline is thus a zone of strong sea-temperature change with depth; it separates the homogeneous warm layer at the surface from the cooler, deeper water.

TOGA = Tropical Ocean and Global Atmosphere

An international research programme carried out between 1985 and 1994 to study the processes linking the tropical ocean, especially the Pacific, to the planet's climate.

Topex-Poseidon

A Franco-American satellite, launched in 1992, to measure changes in **sea-surface topography** level to the nearest centimetre.

Trade winds

A component of the atmospheric circulation around subtropical **anticyclones**. Centred on 15° latitude, these winds blow from the north-east in the northern hemisphere and from the south-east in the southern hemisphere. The trade winds of both hemispheres meet in the **ITCZ** or meteorological equator.

Trophic level

Feeding level in the food web. Plants constitute the lowest level (**primary production**), followed by the herbivores at the next highest level, then by a series of carnivores at the higher levels.

Tropical

Belonging to the tropics: the zone between the Tropics of Cancer (23°27'N) and of Capricorn (23°27'S).

Upwelling

Oceanic phenomenon in which, normally, coastal waters are pushed offshore, towards the open ocean, and are replaced, at the coast, by cooler subsurface water which is richer in **nutrients**. In the open ocean, when the winds or currents provoke the upwelling of subsurface water, the term **divergence** is usually applied.

Vector-borne diseases

Diseases transmitted to human beings by organisms (insects, for example) which inject viruses or parasites into the human body. Malaria, onchoceriasis, trypanosomiasis are diseases transmitted by vectors.

Walker cell (or **circulation)**

Equatorial atmospheric circulation marked by the ascent of warm and humid air (**convection**) over areas of low pressure on the western sides of the oceans in the intertropical zone and its descent over arid high-pressure areas on the eastern sides of the oceans.

Wave

A perturbation, at the surface of, or within, a medium, which moves at a speed depending on the medium's properties.

Zooplankton

Animal plankton comprising organisms whose life cycle occurs entirely in the pelagic domain and others that pass only their larval stage there.